P9-CEO-022

OTHER BOOKS BY FRED ALAN WOLF

THE BODY QUANTUM
*STAR * WAVE*
TAKING THE QUANTUM LEAP
SPACE-TIME AND BEYOND: THE NEW EDITION
(WITH BOB TOBEN)

PARALLEL UNIVERSES

THE SEARCH FOR OTHER WORLDS

FRED ALAN WOLF

SIMON AND SCHUSTER

New York London Toronto Sydney Tokyo

SIMON AND SCHUSTER
Simon & Schuster Building
Rockefeller Center
1230 Avenue of the Americas
New York, New York 10020

Designed by Irving Perkins Associates
Manufactured in the United States of America

10 9 8 7 6 5 4 3 2 1

ISBN 0-671-66091-8

In memory of my loving son,
Michael David Wolf,
who taught me much about the worlds we live in

ACKNOWLEDGMENTS

I wish to thank my wife, Judith Wolf, for all of her efforts in reading and editing this book, as well as all of my others. To Drs. Murray and Susan Brennan, I am grateful for hospitality offered while I worked on the final editing and writing of the manuscript. To Marjorie Valier, I extend thanks for her friendship and hospitality while rewriting the manuscript. I would also like to thank at Simon & Schuster assistant editor Rhonda Johnson for several helpful editing suggestions and my editor, Bob Asahina, for many suggestions and guidance. Thanks are due to Dr. John Cramer and Dr. Susan D'Amato for clarifying some of the ideas. To Bill Whitehead, tragically cut down while in a brilliant career as an editor, I am grateful for first suggesting that I write this book, nearly eight years ago. To the friends and the people of San Miguel de Allende, Mexico, I offer thanks for providing me with a uniquely beautiful village to rework the manuscript. And finally to my beloved son, Michael, killed by a tragic turn of fate, I am grateful for his continued support and love. His memory and love continue to remind me of the spiritual and mystical side of life.

CONTENTS

INTRODUCTION

As Woody Allen once put it, "There is no question that there is an unseen world. The problem is how far is it from midtown and how late is it open?" Since the discoveries of the new physics, the question of the existence of parallel universes—worlds that exist side-by-side along with our own—has taken on renewed interest well beyond mere speculation.

Today, probably more than in any other day, we are facing a revolution in our thinking about the physical universe—the stuff that you and I are made of. This revolution, brought to a head by the discoveries of the new physics, including relativity and quantum mechanics, appears to reach well beyond our previous vision, based as it was on the concept of concrete solid reality. The new physics points in a new and more abstract direction—a direction indicating the need to unify our picture of the world.

The major problem in science today is unification—bringing together a wide disparity of ideas and concepts ranging from the tiniest subatomic matter to the grandest galaxy. Today our knowledge covers a vast spectrum of ideas. And in our attempts to unify those ideas we have discovered great gaps. The science-fictionlike idea that our universe is not alone—that there exist in some mysterious manner alongside of ours (and this needs some explaining) other universes—is the latest concept brought forward by the new physicists in their attempt to unify our knowledge. Without the existence of these other worlds, these gaps of knowledge brought into light by the discoveries of the new physics would remain unbridgeable—incapable of being solved by previous thinking.

When premodern scientific thought about the universe first began with such giants as Kepler, Galileo, Copernicus, and Newton, the universe was imagined to be a gigantic clockwork, with each hand of that clock tipped with a spot marking each planet circling in the heavens around the sun.

Light traveled at infinite or near infinite speed, making every conscious event back here on terra firma always and forever eternally *now* throughout the infinite universe. Five o'clock in Manhattan was also five o'clock on Saturn and on the nearest star. While durations were measurable with clocks, time itself was eternal and unmeasurable. It was infinite and unimaginable. No one could imagine that time *here* and time *there* could have any other relation to each other than the solitary moment of *now.*

And the universe was imagined to be infinite in all directions. There simply was no measure for it. There was no end to space, and to try to think about infinite space was hopeless, a game for fools and poets.

Matter played its game of following exact rules of inertia and movement called *equations of motion,* and nothing in principle was undetermined or, for that matter, left for the imagination. All the universe was a giant machine ticking off throughout all eternity and occupying every corner of an infinite space. Such was our thinking prior to A.D. 1900.

With the twentieth century, ideas of Einstein and the revolution of scientific thinking brought forward by the theories of relativity, much of premodern thinking was changed. Some of the gaps were closed. Space was not as infinite as we had previously thought. It didn't necessarily extend forever, infinite in all directions. Neither was time as inscrutable as thought earlier. Instead time and space joined together and the two became a new concept called *spacetime*. Events were not eternally *now*. For example, two events occurring at different locations could be witnessed by an observer standing in the middle as taking place at the same time. These same events would not be simultaneous for a moving observer. If the observer happened to be moving towards the location of the event on his right, away from the event on his left, he would see the "right" event occurring before the "left" one. And, conversely, if he happened to be moving towards the "left" event, away from the "right" one, he would observe the "left" event occurring earlier than the "right."

Matter was also thought of in a new light. It was produced by the universe itself as a knot in the fabric of spacetime. It *bent* space and it *curved* time. Naturally this changed our vision of the universe's eternality and made it possible to envision just how the universe could have begun. The finite speed of light and the concept of spacetime made it possible to question just what could have occurred when time itself was now imagined to begin, and all space in the universe was imagined to be smaller than the period at the end of this sentence.

However, even with relativity theory, gaps in our knowledge concerning matter and spacetime still exist. Our present models of the beginning of time, called *cosmological* theories, still carry a Newtonian mechanical tinge. They still ring of clockworks, and questions about what happened before the *big bang*—the so-called beginning of everything—paradoxically ring in our heads. And the present models still are grappling with how to bring quantum physics into the beginning of space, time, and matter.

With the discovery of quantum physics—the physics that governs the behavior of atomic and subatomic matter—more gaps in our knowledge were filled. Matter was seen in a very different light. Its properties depended on how it was observed. Thus the actions of observation play a role in the atomic world that was completely unsuspected by the premodern scientists. That role is now suspected to affect even macroscopic matter in subtle ways that could change cosmology and indeed our concept of just what a universe is.

The major problem of bringing together quantum physics and relativity is still with us. We don't know how to do it. We do know that whatever theory that manages it will be quite bizarre for those who still wish a clockwork universe. In this book we will explore one of the most bizarre and promising theories to come from the minds and imagination of today's physicists: that there must be other universes beside our own.

Parallel universe theory began with physicists in the hectic period of the 1950s and 1960s. It appeared as a new way to make concrete and rational some of the bizarre findings of quantum physics and general relativity. These findings aren't comprehensible without a new vision of reality. Instead they appear as problems. Nothing in our previous thinking about the physical world will make these problems go away.

In other words, the existence of parallel universes resolves some old and not too easily solvable paradoxes. However, as you will see soon enough, it introduces a new and apparently paradoxical way of thinking. In essence, parallel universe theory posits the existence of worlds within our technologically extended senses that must connect or relate with our own.

What is a parallel universe? Like an everyday universe it is a region of space and time containing matter, galaxies, stars, planets, and living beings. In other words, a parallel universe is similar and possibly even a duplicate of our own universe. Not only in a parallel universe must there be other human beings, but these may be human beings who are exact duplicates of ourselves and who are connected to ourselves through mechanisms explainable only by using quantum physics concepts.

To see why scientists are now considering parallel universes seriously as a solution to problems in the wide spectrum of thought including modern physics and cosmology we need to explore some new and exciting ideas. Hope of reconciling the ideas contained within this broad spectrum of human knowledge resides in the existence of these other universes—universes that exist side-by-side with our own and even perhaps occupying the same space as our own in some ghostly manner. This spectrum includes quantum physics, unification of new ideas about the universe, relativity, cosmology, a new notion of time, and psychology—or the effects of the human mind on all of this. Consequently, I have divided this book into six parts, each part relating to one of the above. The second part deals with how parallel universes unify our knowledge, and the fifth part deals with how the existence of parallel universes changes our notion of time. Let me comment briefly about the other four parts of the book.

Quantum Physics: Bringing in an Observer

Quantum physics deals with a vast arena of physical phenomena, from subatomic, atomic, molecular, all the way up to modern computer elements such as Josephson junctions, which show quantum behavior on a time and space scale well within the world of human perception. Quantum physics also indicates a new effect—the effect that an observer has upon a physical system. This effect cannot be objectively understood without the existence of *parallel universes.*

Relativity: Relationships—Weird and Wonderful

Relativity, including both the special and the general theories, deals with the relationships between matter, energy, space, and time. It includes many weird and wonderful conceptual ideas, such as gravity being the bending of time and light particles (photons) traveling across the universe without spending any time on their own or going anywhere as viewed from their point of view. A careful look at these classical but non-Newtonian theories indicates that our universe must contain regions of matter that strongly distort the spacetime surrounding them. These regions, called *black holes,* were at first suspected to contain places where the laws of physics would no longer hold. Now we believe that the laws of physics hold everywhere. Consequently these singular regions turn out to be mappable and turn out to be topological *holes* leading to *parallel universes.*

Cosmology: A Search for the Beginning

Cosmology deals with the theory of the early universe—how all the universes began about 15 billion years ago. This theory has gone through a number of important changes. We now realize that earlier theories of cosmology must be wrong because they fail to include quantum physics in their deliberations. By including quantum physics, we find strong evidence for the existence of *parallel universes.*

Psychology: Consciousness and Machine Intelligence

Psychology deals with human consciousness and with problems associated with human behavior and the nature of observation. The *parallel universe* hypothesis enriches the field of psychology. For example, it may help us to understand major disorders, now appearing rampant in our society, such as multiple personality and schizophrenia. I will show how parallel universe theory explains some of the problems dealing with these syndromes.

Psychology also deals with machine intelligence. This book will look into the possibility that parallel universe theory promises a new kind of quantum computer—one that could not exist if parallel universes were not real. This new computer would exhibit the type of intelligence that present-day computers only mimic. Such an intelligence could make the kinds of decisions that we ourselves seem to enjoy. These decisions would be based both on data arising in the past and on data based on the future. In fact, it is my view that parallel universe theory shows that the future can influence the present just as much as the past.

Parallel Universes and Communication with the Future

The fact that the future may play a role in the present is a new prediction of the mathematical laws of quantum physics. If interpreted literally, the mathematical formulas indicate not only how the future enters our present but also how our minds may be able to "sense" the presence of parallel universes.

Are we pressing the mathematical laws of physics too far? "As far as the laws of mathematics refer to reality, they are not certain; and so far as they are certain, they do not refer to reality," wrote Albert Einstein. Einstein was undoubtedly referring to the mathematical laws of quantum physics in that these laws describe only possibilities of reality but never reality itself. Can mathematics describe reality? I believe that the answer is yes, provided we take the new view given us by parallel universe theory. The laboratory of parallel universe experimentation may not lie in a mechanical time machine, à la Jules Verne, but could exist between our ears.

If the parallel universes of relativity are the same as those of quantum theory the possibility exists that parallel universes may be extremely close to us, perhaps only atomic dimensions away but perhaps in a higher dimension of space —an extension into what physicists call *superspace.* Modern neuroscience, through the study of altered states of awareness, schizophrenia, and lucid dreaming, could be indicating the closeness of parallel worlds to our own.

It is in the hope that these radically new and, I believe, quite exciting ideas will turn out to be evidences of truth that I have written *Parallel Universes: The Search for Other Worlds.*

PART ONE

WHAT ARE PARALLEL UNIVERSES?

There is a theory which states that if ever anyone discovers exactly what the Universe is for and why it is here, it will instantly disappear and be replaced by something even more bizarre and inexplicable.

There is another which states that this has already happened.

Douglas Adams
The Restaurant at the End of the Universe

No experiences, ordinary, everyday, usual or unusual, whether impressions, ideas, dreams, visions or memories, strange, bizarre, familiar, weird, psychotic, or sane, are objective facts.

R.D. Laing

No one who has gazed into a full-length mirror has not at some time thought about the right-for-left twisted world that exists on the other side of the looking glass. Indeed Lewis Carroll's Alice has captured our imagination through her magical adventures in a through-the-looking-glass parallel universe consisting of talking playing cards, walruses, ships, and sealing wax. In our imagination we believe that the parallel looking-glass world is as real as our own. Probably you have had the experience of holding up a mirror parallel to another mirror and gazing at the infinite number of images of yourself. Did you wonder, as I sometimes did, if those images were somehow a vision of a truer but stranger reality?

Parallel universes have existed in the fantasies of science-fiction writers probably ever since the genre began. Apparently science-fiction writers have a firm grasp of this rather startling idea. Yet when I try to explain scientifically just what a parallel universe *is,* I find myself stumbling over words. Perhaps the reason is that these universes which are "old hat" to science-fiction writers are rather new to physicists.

Numerous examples of parallel universes from science fiction come to mind. I remember one outstanding *Star Trek* episode. The *Enterprise* had encountered one of those space-warping ionized gas clouds that seem to exist "out there" while in the routine operation of "beaming up" Captain Kirk and some of his crew from a planet below. The ionized gas cloud interfered with the operation of the "transporter," and Captain Kirk and party found themselves on board a parallel *Enterprise,* where the remainder of the crew and the spaceship environment, while much the same as on the old *Enterprise,* were surprisingly different.

Mr. Spock, for example, had sprouted a black pointed beard and, while as logical as his counterpart on the other *Enterprise,* was amazingly evil. In fact, the whole ship that Kirk and party had come upon was as evil as the old *Enterprise* was good in their voyages through the final frontier. Meanwhile, on the "good" old *Enterprise,* an evil Kirk and crew had beamed aboard, and because of their abhorrent behavior the "good" Mr. Spock quickly had them locked up in the "brig."

Both Spocks soon figured out what the trouble was. The *Enterprise,* because of the ionized gas storm, had encountered a parallel universe where a duplicate of the original *Enter-*

prise and crew existed. In fact the duplication was nearly exact in every regard with the exception of some minor details like what was good and what was evil. Neither universe would have even suspected the other's existence if the ion storm hadn't occurred and caused a rip in the spacetime fabric allowing the universes access to each other. However, the parallel Kirks had crossed over into the other's spaceships through the intersection or joining of the two universes. The "good" ship *Enterprise* had an "evil" Kirk locked up in its brig and the "evil" *Enterprise* found itself coping with a "good" Captain Kirk, who soon learned how to pretend he was evil while attempting to right some wrongs. And therein lay a good tale.

Parallel or distorted duplication of what already exists is indeed an important feature of parallel universes according to the way some physicists view them. Accordingly, there are parallel *you*s and *me*s somehow existing in the same space and time that we live in but normally not seen or sensed by us. In these universes, choices and decisions are being made at the very instant you are choosing and deciding. Only the outcomes are different, leading to different but similar worlds.

Legends exist describing *doppelgängers,* or people that are perfect duplicates of other people. These "doubles" are sometimes "space-invaders" coming from a distant galaxy. Who has not looked out at the distant stars and wondered if life did exist "out there"? What would that life be like? Could it have developed like our life on earth? Could it in fact be true that in a far distant galaxy there exists another middle-aged star called Sol and around it whirl nine or ten planets with the third from it appearing as a blue marble when seen from its solitary satellite? Could there be a parallel earth, another planet that is an exact duplicate of ours? Could the forces of the universe create parallel beings like ourselves, and could those beings be in communication with us in some manner that we may only be beginning to detect?

In science fiction there are duplicates of us "out there."

And, in general, these duplicates want to take over and replace us. In a *Twilight Zone* adventure, a woman encounters her parallel self while waiting in a bus station. The double has left her own universe and escaped into this one. The double wants to take over, and indeed manages to do so, while the original woman is locked up in the looney bin.

Aside from there being others like ourselves "out there" in our own universe, there could be beings that are ourselves in the future or in the past that exist side-by-side with us but in a parallel ghostlike manner. In the *Martian Chronicles,*[1] Ray Bradbury describes this other feature of parallel universes. In the story, "August 2002, Night Meeting," a Martian settler, Tomás Gomez, encounters a parallel world on Mars. Just before he takes a drive into the "blue hills," he stops for gas and listens to the old man wiping his windshield. "If you can't take Mars for what she is, you might as well go back to earth. Everything's crazy up here, the soil, the air, the canals, the natives (I never saw any yet, but I hear they're around), the clocks. Even my clock acts funny. Even *time* is crazy up here."

Tomás drives down an ancient highway, and while noticing that he almost could "*touch* time," he encounters a strange jade-green machine like a praying mantis, carrying a Martian "with melted gold for eyes." He cries out "hello" to the Martian, and the Martian calls back "hello" in his own language. But neither understands the other. The Martian reaches over and "touches" Tomás, although Tomás doesn't feel the touch. Now, however, the two can talk to each other in the same language. When they try to touch each other again to shake hands, each hand passes through the other's as if the other didn't exist. Yet they see each other, but to each the other is not solid.

They figure out that they are in parallel universes that just happened to overlap now and there. They each feel their own solidity. They each believe in their own reality and sense the other as a ghost. They try to figure out how their worlds are touching without touching, but to no avail. The Martian, looking over the Martian landscape of his parallel world, sees a beautiful city with wondrous things. Tomás, looking over his world seeing the desolate and deserted ruins of a city, tells the Martian that the city is in ruins—over thousands of years old. "The canals are empty right there," shouts Tomás. "The canals are full of lavender wine!" says the Martian.

They realize that their meeting has something to do with time, but can't decide who is living in the future and who in the past. They part, each returning to his respective world and each thinking that the other world was a strange dream.

Although strange and otherworldly, these stories today have a ring of truth to them. As we shall see in what follows, the story of parallel universes as put together by science fact is a strange tale indeed.

Chapter 1

HOW QUANTUM PHYSICS PREDICTS THE EXISTENCE OF PARALLEL UNIVERSES

Quantum physics or quantum mechanics, which is the same thing, is a strange business. It deals with the behavior of matter and energy, particularly with how matter and energy interact on a very tiny scale—the scale of atoms, molecules, and particles that exist inside these small objects. Now atoms are quite tiny and usually we, in our everyday lives, need not concern ourselves with them. An atom is so small that if my thumb inflated, like a balloon, to the size of the planet we all inhabit, one tiny atom of hydrogen contained in one tiny molecule of water which makes up a tiny droplet of perspiration on that thumb would then be quite visible, assuming that it too inflated the same way as my thumb. It would then be as big as my real thumb. In other words, my thumb compares in size to the whole planet earth, as a single atom of hydrogen compares to my thumb.

Even an atom is a large thing when compared to a subatomic particle. The word *subatomic* means smaller than an atom or capable of existing within an atom. One very important subatomic particle is the electron. Electrons live inside of atoms. But just how they manage that was not really understood until the invention of quantum mechanics. It appeared that if an electron were to follow the laws of physics that existed before the discovery of quantum theory—the so-called classical physics invented by Sir Isaac Newton and James Clerk Maxwell—it would never be able to leave the atom. Nor would an electron ever be able to emit radiation, as it does when a light bulb is turned on.

Electrons leave atoms all the time. They do so easily when atoms are grouped together, such as in a solid object like a copper wire. Indeed without electrons flowing easily through copper there would be no such thing as electricity. The reason for this is that electrons obey quantum physical laws. They are electrically charged and are attracted to the centers of the atoms they occupy. These centers are the nuclei of the atoms, and they too are electrically charged. Electrons have a negative charge and the nuclei have a positive charge. Consequently, the electrons are attracted to the nuclei so strongly that according to classical physics every electron in every atom would be gobbled up by every nucleus in about one hundred microseconds (a microsecond is a millionth part of a second).

If this were to happen, all atoms would drastically shrink in size, and all materials, which are made of atoms, would also shrink in size, undergoing a horrible contraction. To gain some idea of this horror, consider that a real atom is quite a bit larger than its nucleus. If an atom were as big as a football stadium, its nucleus would be about the size of a Ping-Pong ball situated on the fifty-yard line. So if an electron inside an atom were to follow the laws of classical physics, it would be gobbled up by the nucleus of the atom, and the atom would then shrink in size by the same ratio that a football stadium compares in size to a Ping-Pong ball.

Everything would shrink in the same way, since everything is made of atoms. Our planet would shrink in size so that it would be about one hundred feet longer than a football stadium in diameter! A six-foot-tall person would be smaller than a blood cell.

Of course, this doesn't happen. But to explain why it doesn't, physicists needed a new theory—a new vision of the atom. That came from the new physics called quantum mechanics. However, this new vision also contained some very strange ideas.

The New Physics Had Strange Ideas

The first curious idea was that tiny objects could not move as large objects apparently seem to move in our everyday world. Atom-sized objects moved in jerky leaps from one place to another without going in between. These movements were called quantum jumps, and no one could predict with any certainty just where an object was going to be after appearing somewhere.

The second idea was that tiny objects could not exist objectively independent of the observers of those objects. Somehow in the very act of observation, these tiny objects took on characteristics that could not have been present before they were observed. By looking for one feature of an object, one completely altered the object's other features in unpredictable ways. Thus what one chose to examine altered what existed.

The third idea was that there had to be a new order in the universe, despite the apparent disorder presented by the first two ideas. This order was not the order expected using the old or classical physics—instead it was an order that involved us! It involved our minds in a way that we could have never expected using the old physics. This order said that we were in control of possibilities but not actualities. We could predict where and when something was likely to occur but not where and when it would occur.

All of these ideas arose in a very short period of time between the turn of the twentieth century and the present. All of them are still true today.

It was to make sense of these ideas that parallel universes were first conceived of by Hugh Everett III, a physicist at Princeton University in 1957. Before we look at Everett's ideas and see how they made sense of the seemingly nonsensical quantum world, we will take a closer look at the experimental facts of quantum physics.

Chapter 2

THE PENULTIMATE EXPERIMENT: SHOOTING THROUGH DOUBLE SLITS

No better example exists illustrating quantum weirdness and indicating why parallel universes manifest than the double-slit experiment. In this experiment, a stream of subatomic particles, electrons, atoms, or even light itself is directed toward a screen, much like a beam of light is projected from a movie projector toward the silver screen in a movie house. A barrier is placed between the screen and the projector. It contains a pair of parallel slits that are closely separated. Each slit can be opened or closed independently. The intensity of the stream of particles is highly reduced until only one particle passes toward the slits at any given time. Thus each particle then encounters the slits, passes through, and registers on the screen. Or each particle hits the barrier and is absorbed. Even if the stream is well aimed, no one can predict just where a particle will land.

The screen is sensitive. Every time a particle manages to

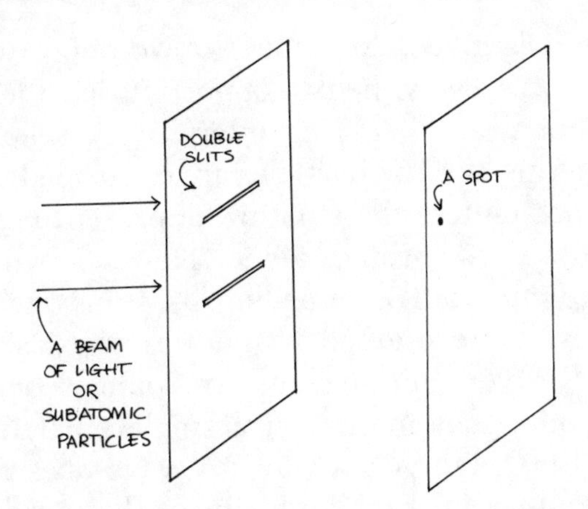

pass through the slits in the barrier it eventually leaves a tiny spot on the screen. In this way, so one would think, each particle must pass through one slit or the other in order to reach the screen.

Yet the amazing thing is that if you close down one of the slits more particles reach certain places on the screen than if you leave both slits open. There is really no way to understand this fact if you think that the stream is composed, as I said it was, of single tiny particles. How does a single particle know if you have two slits or only one slit open? Since each particle apparently has a choice of which slit to pass through, each should have twice the opportunity of reaching any point on the screen. This means that with both slits open, the particles should reach any point on the screen with a greater frequency. In other words, with two slits open, twice as many particles should have reached any point on the screen.

Yet, as soon as you close down one of the slits, denying the particles any choice, they somehow manage to reach places on the screen that they never manage to reach when both slits are open.

How can anyone explain this bizarre behavior? Why do the particles avoid making spots on certain areas of the screen when both slits are open? No ordinary commonsense picture of a tiny object explains the weird behavior exhibited by a particle when it is given two chances, two opportunities, to make a fact. In some way the two opportunities, the two possibilities, affect one another. They interfere with each other.

But how can this be? The particles are controlled so that no particle ever encounters the slits in the presence of another particle. Each passes through alone. Thus no two or more particles bump into each other as they pass through. Can this interference result be explained by quantum physics?

The answer is yes, but the answer changes our way of thinking. Suppose that instead of a single particle passing through the slits, there was a wave. A wave doesn't behave like a particle. It could reach both slits at the same time. No particle could do that. The wave could then break up into two waves, one passing through each slit. This happens all the time when a real wave, such as a common onshore ocean wave, comes to two openings such as the space between three parallel piers. In this manner, the two waves in the stream would travel separate paths and reach the screen. Since the waves would travel different paths to reach the screen, the waves could interfere with each other.

Waves are made of moving, rolling hills and valleys. If the point on the screen where the waves hit is a point where a hill of one of the waves meets with a valley of the other, the waves would cancel out there. That would explain why there were places on the screen where no spots appeared when both slits were open. Close down one of the slits, and the wave would not be broken up into two parts. All of the wave would reach the screen after passing through a single slit.

If one uses a wave description, this turns out to solve the problem. Indeed quantum physics was at first called wave mechanics, because a wave description solved this and other problems that occurred whenever a subatomic object was faced with two or more possibilities. The possibilities always interfered with each other as if the object was in some way a wave. So that meant that subatomic matter was really composed of waves. There were no particles, after all, in the stream. The stream was made of waves.

But that isn't correct either. When the waves arrive at the screen, they do not land everywhere on the screen, like a wave reaching a beach. Instead, when the waves arrive they somehow "hit the beach" in a series of pointlike spots. Real ocean waves don't do that. They wash up on the beach at many places at the same time. And in other experiments, called particle-scattering experiments involving subatomic and atomic particles traveling through space, the same thing turns out to be true. The final outcome is always that a particle leaves a track, a spot somewhere, but travels through space as if it were a wave. Thus "waves" are still "particles?" after all.

This behavior of particles when confronted with two or more possibilities is called the *wave-particle duality.* But calling it by a name doesn't solve the problem. We are faced with a mystery.

The Mystery Persists

The enigma remains with us today. No one knows why subatomic matter behaves this way. We do know that the rules of quantum physics work. With its rules we can explain many, many previously unexplainable facts of physical life. Yet, despite its enormous practical successes (quantum theory correctly predicts the behavior of such things as lasers, microchips, photocells, nuclear reactors, long-range deep-space communication devices, many types of solid-state inventions, transistors, and materials at very low temperatures, to mention just a few), quantum theory is still so contrary to intuition that, even after eighty years since its inception, most experts do not agree what to make of it.

The Wave Was Not Real

At first physicists believed that the wave was not real, in spite of its having a real effect in the world. They believed that the wave somehow existed only in our minds as a means to keep track of experiments. It was a "wave of possibilities." (This wave, by the way, is called the *quantum wave function* or *probability wave.*) Thus in the double-slit experiment, there was really only one particle present at any time near the slits, and the wave represented the different possibilities that the particle had. In this way, one explained that only the possibility or, in the language of quantum physics, the probability, of that particle moving through the slits was determined. One could not predict where a particle would go. All one could do was predict where a particle was likely to go. When there were two slits open, the particle had two ways to go.

The wave enabled physicists to calculate very nicely the probabilities of particles doing things. With one slit open, the probability was changed from the situation where both slits were open. Although the double slits did pose the interesting problem that somehow these probabilities affected each other, since the particle in passing through the slit arrangement somehow had to pass through *both* at the same time, it was easier to think that the wave wasn't real and that only a single particle was there at any moment.

Why Parallel Universes Were Invented

This is why parallel universes came to be considered as a serious way to explain this fundamental mystery. In 1957, Hugh Everett III, a graduate student at Princeton University studying under the highly regarded physicist John Archibald Wheeler, came up with the rather strange notion[2] that we should take quantum mechanics seriously. If it says that two alternatives can interfere with each other, then somehow those alternatives must both exist simultaneously. If possibilities can affect each other, if two or more probabilities can somehow "add up," then somehow these possibilities must really exist somewhere. But where?

Apparently, the wave of possibilities was composed of a

number of particles—with each particle really existing somehow in a separate world. In this way, having separate or parallel worlds, only one particle would ever be found in any one world. That would explain why only a single spot was discovered after the particle passed through the slits. No one knew exactly where that spot would be found, but nevertheless it would be found somewhere.

In the double-slit experiment, only two worlds were necessary. In one world, the particle passed through one slit. In the other world, the particle passed through the other. The two worlds would exist side-by-side, completely separate from each other until the particle reached the screen. Then the two worlds would overlap or merge. Why would the worlds merge after splitting apart? The answer was even stranger, although it did satisfy the inquiry.

The reason was self-consistency. This was the only way in which the interference could be explained and still have only one particle. The universe itself was continually doing this splitting and merging each and every time that anything interacted with anything else. Each split was necessary to produce the wave behavior and each merger was necessary to produce the particle.

In this view, the wave represents not possibilities or likelihoods but realities—an infinite number of them, if necessary. The wave is composed of particles in parallel worlds. When the particle strikes the slits, the world, indeed the whole universe, containing it splits into two (and, in general, a multiplicity of) mutually unobservable but equally real worlds. In each one of these worlds a particle passes through one slit. When the particle strikes the screen the worlds coalesce, forming a single world once again. Measurement gives a definite result. And all is well because it ends well. Although this proposal leads to a bizarre world view, it may be the most satisfying answer yet devised.

Chapter 3

RIDING THE WAVE THROUGH PARALLEL WORLDS

In the parallel universe model, the wave of possibilities is something that at first sight appears magical. Let me describe its properties. First of all, it is a wave and it can be imagined to do all the things that waves do. It undulates, vibrates, and moves through space and time. The only thing weird about it is the space and time that it moves through. When the wave encounters a situation that is logically impossible in a single world—a single spacetime region—the wave does something analogous to the way an ocean wave behaves when it encounters two or more spaces between piers jutting out in the ocean space. The wave splits apart. One part goes off running after one possibility and the other part follows the other possibilities. Each possibility is a reality in a different world.

For example, a flipped coin falls heads. The same coin falls tails in a parallel world. The wave that represents the coin contains both possibilities. It contains even more possibilities for the coin. It contains the possibility that the coin could land on its side. It contains the possibility that the coin could suddenly absorb a great deal of energy and boil off into separate atoms. It contains the possibility that the coin could vanish right before your eyes and then reappear on the other side of the room. Anything that can happen, even though extremely remote, is contained in the wave.

The different possibilities are limited only by the imagination of the experimenter. If a certain event is more likely to occur somewhere rather than somewhere else, the wave that represents the different possibilities will be more concentrated, higher in intensity, in the neighborhood of that event. The wave takes advantage of each opening possibility for it to flow into. Just as an ocean wave splits into parts flowing into different channels, eddies, and God knows whatever, the wave of possibilities takes advantage of every opening, every nook and cranny, every smidgen of possibility that an experimenter can devise.

If those possibilities are changed by an observer or experimenter, the wave changes accordingly. Just as a wave can be split into parts and then recombined after passing through or around obstacles, the wave of possibilities can be split apart and can be merged too. Always, like the great oceans that surround us, the wave persists.

What is perfectly simple from a wave point of view is perfectly bizarre when one puts oneself into the picture or when one attempts to describe what is going on at the level of atoms and molecules. At this level, the wave is totally invisible. The different possibilities are not waves spilling over jetties and around barriers or piers, they are realities in different worlds. Each world appears and disappears—recombining back into one world—each time a subatomic particle interacts with something.

The world seems flimsy when seen from the wave of possibilities—parallel worlds—view. Its strength comes from the fact that many, many possibilities combine, creating what appears to be one gigantic reality. But this reality is just a combination of other flimsier realities. And these realities are continuing to split apart and rejoin every time something called an *interaction* takes place.

In a wonderful imaginative story by Jorge Borges, called "The Garden of the Forking Paths," the outrageous world of parallel worlds is described as

> *an infinite series of times, in a dizzily growing, ever spreading network of diverging, converging and parallel times. This web of time—the strands of which approach one another, bifurcate, intersect or ignore each other through the centuries—embraces every possibility. We do not exist in most of them. In some you exist and not I, while in others I do, and you do not, and in yet others both of us exist. In this one, which chance had favored me, you have come to my gate. In another, you, crossing the garden, have found me dead. In yet another, I say these very same words, but am an error, a phantom.*[3]

A Dizzy Atom in an Infinity of Worlds

To see how the parallel universe idea explains the world we experience when dealing with subatomic matter, consider the following picture we have of the simplest atom in existence, the hydrogen atom. A hydrogen atom consists of a single nucleus that contains a positive electrical charge, called a proton, and, somehow surrounding it, a single subatomic particle called the electron. The electron has a negative electrical charge. The proton and the electron are held together by the electrical force of attraction between them.

But that doesn't explain the atom's properties by a long shot. How do the electron and the proton arrange themselves inside the atom? One picture that persisted for a long time before the quantum theory was fully developed was that the electron ran around the proton in an orbit much as the earth runs around the sun. But that turned out to be incorrect, because the force of attraction pulling on the electron would eventually cause it to accelerate and radiate away all of its energy. Consequently, as I mentioned earlier, the electron would simply take a dive-bomb into the nucleus, and the atom would diminish in size.

The present quantum physical picture of the electron is that in order for it not to fall into the nucleus, it exists in the form of a cloud, not a particle. The cloud is made up of separate electrons, each occupying a different position in a separate world or universe. As long as no one tries to locate it, these separate universes overlay, creating a single world. In this world the electron is a cloud in space, and this cloud actually gives the atom its shape in the world and provides it with a stable and defined energy.

With the change in shape of the cloud, the atom is observed to have a different energy—each energy corresponding to a specific cloud shape. This cloud, however, is difficult to understand, because whenever an electron is observed in any experiment, the electron does not exhibit any cloud form at all. It appears as a particle—a spot with no apparent spread in space. And when that happens, the atom no longer has a defined energy.

The disparity between the observation of the energy of the atom and the precise determination of the location of an electron in it required a new principle. This was called the *principle of indeterminism* or the *Heisenberg uncertainty principle* (UP).

Heisenberg's uncertainty principle is a concept that reflects the inability to predict the future based on the past or based on the present. It arose from the ideas and thoughts first stated by Werner Heisenberg around 1926 or 1927. It is the cornerstone of quantum physics and provides an understanding of why the world is made of events that cannot be related entirely in terms of cause and effect. It is at the root of all physical matter and may manifest for human beings as doubt and insecurity. If this is so, once it is fully understood, it could create a condition of enlightenment in which the world is seen as an illusion and as a product of mind or consciousness.

It says in essence that it is impossible to know both the location of a subatomic particle and its energy or momentum simultaneously. Either the energy or its location is determinable at the expense of the other. There is no way to explain this disparity using any commonsense idea.

The Quantum Magic Lantern Theater

To visualize how an electron in an atom appears in parallel universes, imagine entering a magic theater where a slide show about the atom is about to be presented. Picture each of these separate electron worlds as captured on a thin transparent filmslide, each slide corresponding to a universe in which the electron in the atom has a definite position. And imagine a gigantic dark theater, containing an infinite number of slide projectors, one electron-position slide in each projector. Now imagine that you were to witness projected on a single screen the image from all of these slides. Each slide contains a picture of the electron in each possible position in the atom. Yet this is not what you would witness.

You would be seeing the overlap of an infinite number of parallel universes, and that overlap would appear to you, not as an electron in an atom with a particular position—a single spot pointing to the electron's existence—but as an atom with an electronic cloud surrounding it. And that electronic cloud would persist, never changing as time marched on. This cloud would show the atom in a stable pattern of energy. Now a single slide contains just a dot, and if the projectionist rapidly flicked from slide to slide you would see a spot on the screen jumping from point to point. In this manner, the electron is said to make quantum jumps from place to place; however, the energy of the electron at the moment of each jump is indeterminate and hence unpredictable.

By instructing the projectionist just which combination of slides to show, in other words, depending on the combinations of all those position-universe slides, any one of an infinite number of unique patterns will emerge on the screen. Each pattern corresponds to a unique electronic energy and is called an *energy state*. When the lowest energy or ground state pattern emerges we say the electron is in the lowest energy state, but it does not have a unique location.

Suppose that a photographer comes along and takes snaps of each possible cloud pattern—each possible energy state of the atom as projected on that screen. Suppose that the theater's dark room develops those snaps, making slides. If you call for a new slide show and ask the projectionist to put these *electronic energy-cloud* slides into the projectors, you would in essence change the show. Each slide contains a picture of an atomic cloud. If you were to witness projected on a single screen the image from all of these slides, you would not see an overlap of clouds. Although each slide corresponds to a universe in which the atom has a definite energy, this is not at all what you would see.

You would be witnessing the overlap of an infinite number of parallel universes; in each universe-slide the atom would have a stable energy. But that overlap would appear to you, not as an atom with a stable energy, but as a single spot pointing to the electron's existence. As time marched on that spot would quantum jump from place to place on the screen, just as before, when the position-slides were changed, provided you were there to observe it. When you weren't looking directly—kind of slyly sneaking a look out of the sides of your eyes—you might catch a glimpse of the ghostly cloud patterns. But any direct viewing would only catch a spot.

Thus, in this noncommonsense manner we can visualize the atom and understand the UP's rule. We also can see how any universe is composed of an infinite number of other parallel universes.

The Electron Exists in Parallel Universes

The parallel worlds idea says that the electron exists as a single particle all of the time. Each position it takes, however, does not exist in a single universe. Instead it exists in a dizzying overlap of an infinite number of parallel worlds. This overlap appears to us, provided we don't actually pull the electron out of the atom, as a single world. Thus the world we experience of electronic clouds in atoms is an overlap of parallel worlds in which the electrons exist as particles. In other words, the world of atoms we experience is itself a composite of overlapping parallel worlds.

This world of atoms persists in this overlapped condition until something comes along and disturbs the overlap—say, a large electrical force that pulls on the cloud causing it to split apart. When that happens, all of the overlapping worlds suddenly fly apart. In each now separate world, the electron appears as the single particle it is. But until that disturbance takes place, the cloud persists, the worlds overlap, and the atom simultaneously inhabits all of the worlds as if the sum total of them were one single world.

We see the evidence of this when we split the atom, causing the electron to be pulled apart. If we repeat the experiment many, many times, the electron is never observed to fly away in the same direction each time, no matter how carefully we perform the experiment. This is a clue that what we are seeing is a different world each time.

Chapter
4

A WAVE PASSES THROUGH COPENHAGEN

However, this isn't the only possible way to explain the bizarre behavior of atoms and other tiny objects. The parallel worlds idea, in fact, is not widely accepted because it appears to posit such a strange universe. No one yet has seen a parallel world while living in this one. At least, so we think.* The most accepted way today to interpret what goes on when dealing with the subatomic world is called the Copenhagen interpretation. It was invented by Niels Bohr, who had set up his famous Copenhagen Institute in Copenhagen, Denmark, around 1927, and is generally accepted by most physicists to be the explanation of quantum weirdness.

Niels Bohr, who is considered, in good quantum physical

*In Part Six, we will consider several possibilities for detecting parallel worlds while living in this one.

paradoxical language, the father and the mother of quantum physics, came up with his interpretation in order to explain how atomic and subatomic objects no longer have the same attributes as pictured by Newtonian physics. In Newtonian or classical physics any object is believed to possess two kinds of attributes simultaneously. For example, a large object is pictured as having both a location and a momentum at each and every moment. Together, the momentum and the location of the object tell us where the object is and where it is going or, in other words, we know both the position of it and its path or trajectory as it makes its way from one place to another. In quantum physics dealing with tiny objects like atoms, electrons, and other subatomic particles, this is no longer possible. Such objects cannot be observed with both positions and momenta simultaneously.

Bohr's interpretation gave a reason for this behavior. He said that tiny objects aren't what large objects seem to be. A large object follows Newton's laws. Its path and location are simultaneously observable. But atom-sized objects are disturbed by any attempt to observe them. For example, if one carefully performs an experiment to observe an electron's position, that experiment necessarily blurs its path. Conversely, any experiment to determine the electron's momentum makes it impossible to determine its location.

Bohr believed that this wasn't due to any ineptness on the part of the experimenter. Rather, it was due to the inevitable consequence that eventually a large object, such as a machine, recording device, or human being, had to observe a tiny object, such as an electron or atom. The large object followed Newton's laws while the atom-sized object didn't. Since any information about the tiny object had to be obtained by a large object, all it could do was disturb the little particle with unpredictable results. These disturbances were viewed as *quantum jumps*—tiny jitterbugging taking place on an atomic scale. The experiment could never really tell exactly what was happening.

Thus Bohr considered it meaningless to picture what an

atom-sized object really looked like. For how could anyone ever know? The best one can say about tiny objects is to describe what can be observed about them. No more and no less. If you can't observe a tiny object's path and its location at the same time, it is worthless to assume that it has these attributes simultaneously.

Now Bohr's views have come to be the accepted interpretation of quantum weirdness. It's as if a religious canon had been handed down to Bohr's followers: "What cannot be seen should not be discussed."

However, this interpretation is really beyond the laws of quantum physics. It is actually an added assumption, and a very strong one at that. Thus even today, the question of the meaning of quantum physics is certainly not a dead one.

During the writing of this book, I attended a meeting[4] of top physicists held at the New York Academy of Sciences. The subject of the four-day-long conference was "New Techniques and Ideas in Quantum Measurement Theory." By measurement, these physicists meant the effect that an observer of an atom or subatomic system has on the system. As we will see shortly, measurement is a problem related to the appearance of parallel universes.

The conference was dedicated to physics professor Eugene Wigner—a forerunner in the business of quantum physics—in the hope that the presentation of new ideas and areas of thought would be able to stimulate further progress toward understanding the profound structure and possible limits of quantum theory. A meeting like this was quite welcomed. Nothing like it had occurred since the early days when Einstein, Bohr, and other giants of the new physics first began worrying about it.

Indeed such a meeting was a delight for all in attendance. Ex-physics student and senior editor for *Science News,* Dietrick E. Thomsen,[5] attended and remarked that a feeling of excitement ran through the meeting akin to what he had felt earlier when, as a student, his professor announced the discovery of the neutrino, a new fundamental particle.

The big problem in quantum physics is: what happens when a measurement of a physical system takes place? According to the quantum gospel, also known as the Copenhagen interpretation,[6] whenever an observation occurs the physical system undergoes a sudden change in its physical properties. Before the measurement, the system is described by a composite of possible physical states. After the measurement, the system is found in one of those states. The big question is how does the measurement or simple act of observation suddenly change the system? Up to 1986, the question appeared to be one that had to be decided by theory.

In my days as a graduate student in theoretical physics at UCLA (after receiving my B.S. degree in physics at the University of Illinois), there were many mysteries involving the actions of an observer. Such mythological beasts as Schrödinger's cat, Wigner's friend, and the Einstein–Podolsky–Rosen paradox were standard paradoxes exhibiting this serious problem in quantum physics. Yet as outstanding as these insights into the mystery of observation were, no one had ever attempted to devise a way to measure any one of them.

Schrödinger's cat refers to a poor pussy that has been locked up in a box containing a Rube Goldberg device that will or will not emit cyanide gas depending on the outcome of a single quantum event—the radioactive discharge of an atom. The paradox is: Suppose that the cat is in the box for a period of time wherein the probability is fifty percent that the atom has discharged. If no one looks in the box, is the cat dead or alive? No answer comes from Copenhagen. In the parallel universes model, the cat is both alive and dead in separate but equal worlds.

Wigner's friend refers to the Schrödinger cat paradox. Suppose the friend who holds the cage containing the cat decides to look in. He will undoubtedly find a live or a dead cat. But suppose a professor named Wigner has the friend and the caged cat in a closed room. If the professor doesn't look in on the friend, even though the friend has looked at the cat, is the friend in a happy state of mind upon seeing a live cat, or a sad state of mind upon seeing a dead cat? According to quantum rules written in Copenhagen, until the professor looks, the friend's state cannot be decided. But according to the parallel universes thesis, the friend and the cat exist in two editions.

The EPR paradox, named after physicists Albert Einstein, Boris Podolsky, and Nathan Rosen, deals with a measurement performed on one part of a physical system while the other part, which had previously been connected to it, is left alone. What happens is the measured part (part one) instantly affects the unmeasured part (part two) at the moment of measurement, even though there is no longer any connection between the parts. If, for example, the measurer determines the energy of part one, the energy of part two is instantly determined. If, on the other hand, the observer measures the position of part one, the position of part two is instantly determined.

This ability for an observation which occurs at one location to instantly provide an object that is at another with a physical attribute isn't really considered by Copenhagen followers to be a problem. It isn't considered not to be a problem either. Instead, Copenhagen followers reading from the Bohr quantum Bible simply raise their voices on high and say in sermonlike cryptic tones, "If it is observable, there is no paradox. If it is not observable, why talk about it?"

But, commonsense raises its head and says, "Hold on Jack, I mean, Niels. How can information get around so fast, specifically, instantly?"

The accepted manner in which information moves from one place to another is through the action of a carrier of that information such as an electromagnetic or sound wave. But no carrier is available in the EPR paradox. This doesn't bother the students of the Copenhagen school because they redefine information as that which can only travel from one place to another via such a carrier. The instant communication between separated events is not information, they would proclaim.

In the parallel worlds model there is no paradox, for whatever is measured about the first part is already determined in the second part. All possibilities of energy or position measurements exist in separate worlds.

All of these measurement paradoxes pointed to what I call the trouble with the quantum as seen by the conventional school. That trouble can be stated, "How is it that something is?" How does reality as we experience it arise? How, in other words, does a physical fact register as a physical fact? How does one go from a mere theoretical description of nature, as, for example, provided by quantum theory, to the real tangible world we all experience as "out there"?

Trouble in the New Physics

Nearly all physicists find that quantum physics is troublesome because of the problem associated with the action taken when an observation occurs. Even though relativity is troublesome, at least it fits a certain consistent picture. Matter and energy are still objective facts of life. They exist even if they are bound up inexorably with space and time. Two observers may disagree on the time interval* between two events, but each knows how to determine the other's viewpoint. In other words, relativity is still a classical science. Determinism† does not get thrown out like the baby with the classical bath water.

But observers in quantum physics do not see events deterministically. Not only do they not agree with each other but

*See Part Three for more discussion of this point.

†Determinism is the philosophy that says what is depends on what was. What will be depends on what is. All are connected across time by laws of physics that allow one to determine what happens based on what happened, and what will be based on what is happening now.

they are inextricably bound to what they observe. The observer disturbs the universe in the quantum world. She can do this in two possible ways.

The Observer Participates Whether She Wants to or Not

In the Copenhagen view, nothing like a parallel world exists. Instead the wave of all possibilities undergoes a sudden change the instant anything physical is observed. This is called the *collapse of the wave function.* It can be imagined to occur just like a pricked balloon suddenly collapsing. The observer is responsible for the *collapse of the wave function.** She looks at the system, and the system suddenly quantum leaps into one of the possible states. The observer stands outside and, when she looks in, the system observed suddenly takes on a physical value.

In the parallel universe idea, the observer is part of what he or she chooses to measure. No collapse of the wave takes place. The wave just keeps truckin', but now the observer becomes part of the wave.

In quantum physics, without bringing in parallel worlds, there is no way to determine this sudden collapse. In fact, quantum physics does not predict that any such collapse even takes place. This is the essence of the Copenhagen position—an added assumption of sudden collapse. Physicists who believe in the Copenhagen view "know" that such a collapse occurs because the system appears to take a physical value, and they can see that it does.

However, according to quantum rules, no collapse occurs. Instead the observer, in the act of observing, becomes involved in the system. Whenever the observer is part of the

*The sudden change in the quantum wave function when an observation takes place. Since the wave function represents the probability of observing an event, the collapse means that the probability has changed from less than certain to certainty. This is also known as the *observer effect.*

system being observed, the whole—consisting of both the observer and the system—evolves in time in a completely correlated and consistent manner. The only trouble is that the whole system must evolve into a myriad of splitting and merging parallel universes.

Suppose the quality of the object being observed is its color. In the world of everyday experience, we just look and the color is seen. If the object is capable of taking on different colors and it is seen to be black, then it is black. If the object is seen to be white, then it is white.

But how is this to be explained using quantum physics? If no collapse occurs, the object and the observer are tied together. All that exists is possibilities. In one possibility, the object is white and the observer sees white, and in the other possibility the object is black and the observer sees black.

Thus if a quantum system can evolve to any one of a set of possible states, and according to quantum physics it evolves to all possible states simultaneously, the observer of that system must also evolve to all possible recognitions of those system states. The observer becomes part of the system she observes.

This tie of the observer to the thing being observed persists until the observer sees the object is black or sees it is white. Then a miracle occurs, according to Bohr and his school of thinking. The measurement problem is over and the wave of possibilities is reduced to one. This Copenhagen school or interpretation is, as I mentioned, the "conventional" interpretation of quantum physics.

Physicist Bryce DeWitt, a leading exponent of the parallel worlds hypothesis, points out that if a poll of physicists were conducted, the majority would profess membership in the Copenhagen or conventional camp, just as most Americans would claim belief in the Bill of Rights, whether they had ever read it or not. DeWitt believes that most of these physicists do not really know that they often change the rules of quantum physics in their interpretations believing that they are subscribing to the conventionalist's camp.

Thus the Copenhagen school is just too open, admitting anybody to its classes. The Copenhagen school postulates an immediate "split-and-collapse" of the wave function as soon as the observer is connected to the system. This idea is fraught with misunderstanding and confusion. One cannot say into which state the collapse takes place. All one can do is assign a statistical weight to each state according to the relative "height" of the peak of the wave representing that state. Thus the wave with the highest peak spread over the most space grabs the most probability and therefore is the best candidate for a real "happening."

The other possibilities still exist. But then comes the collapse. This collapse is not found in the mathematical description, called the Schrödinger equation, that provides the time evolution of the wave. The collapse is a consequence of adding an interpretation to the physics after the fact. And the major reason this must be done is that no one quite knows what to do with the quantum wave function—the wave of all possibilities. Is it real or is it just a fantasy?

Chapter 5

THE MEANING OF A QUANTUM WAVE FUNCTION

A quantum wave is not quite the same thing as a real wave, although they share many attributes. The quantum wave was first conceived by French physicist Louis de Broglie, in 1923, in his attempt to explain why electrons in atoms did not themselves fall into the nucleus and radiate away all of their energy. De Broglie was attempting to explain the mysterious manner in which electrons were thought to behave when confined to the boundaries of an atom.

Niels Bohr postulated that atomic electrons would not radiate energy when they were in certain constrained types of circular motion later called Bohr orbits. De Broglie added that there was a wave of something carrying the electron in one of its undulations wherever a Bohr orbit existed. Later on, Erwin Schrödinger developed a mathematical representation of the wave. He believed that the wave was electromagnetic in character, much like a radio or television wave. His equation is now de rigueur in a course in quantum physics.

The Wave Was Invisible

It turned out that the wave was not electromagnetic in character at all and, in fact, was never observed. Instead the wave was interpreted, by German physicist Max Born, to be a wave of probability, not a real wave at all. It described the probable location of an object in space but never its actual location.

When Schrödinger's mathematical description was developed for two- and then three-electron atoms, it was able to successfully predict their general atomic structure and the character of the radiation those atoms did emit when they gave out radiation. It successfully explained how the electrons were able to give off radiant energy and not be gobbled up by the nucleus. This was not possible using Newtonian physics or even the new model of the atom that Niels Bohr conceived. The new wave did explain the atom remarkably well. The de Broglie–Schrödinger wave description became so successful that every physicist was using it to describe atomic behavior.

In fact the de Broglie–Schrödinger electron-wave was so accepted by everyone that they literally abandoned the idea that the electron existed in an atom as a particle. Again, we only have to look back on Copenhagen to see why. "That which cannot be observed cannot be discussed," goes the dogma. Since no one can ever really see the electron existing as a particle inside an atom—for whatever attempt made to make that observation will so disrupt the atom that the electron will no longer be in the atom—it is needless baggage to carry the electron piggyback on the wave any longer. Let the little "bugger" go, cry the indoctrinated.

As time marched on, the wave became the nouveau mode

and, like all fashion, everything else soon became forgotten. Perhaps forgotten, but not gone according to those who did not side with Bohr. De Broglie certainly did not want the electron-particle to vanish in a puff of wave. To him the wave simply carried the particle. But to the fashionable the particle was gone and the new wave became as much a cornerstone of physics as the Heisenberg uncertainty principle.*

The Emperor's New Clothes: The Quantum Wave Function

The wave was useful for describing atoms. But it did not give a picture of atoms that one was used to with Newtonian physics. The electrons in the atoms were viewed, not as tiny point particles, but as being made from the wave itself in some yet mysterious way.

In the de Broglie view (later taken up by David Bohm, Basil Hiley, Jean-Pierre Vigier, and others[7]), the electron was still a solid little particle buffeted about by the de Broglie wave. This view has never been experimentally testable and is not popular because it posits the extra assumption that the wave and the particle are separate and in some cases separable (which also is untestable so far†).

*See Chapter 3 for a description of Heisenberg's uncertainty principle.

†Jean-Pierre Vigier and his co-workers were quite evident at the New York quantum measurements conference I mentioned earlier. Vigier enjoys being at the center of controversies. He told me that he was once an adviser to Ho Chi Minh during the French-Vietnam crisis! Having had past connections with the Communist party did not make him very popular in the United States, I am sure. One incident of our meeting came to mind. His suitcase containing the slides for his talk was due to arrive from Paris, of course, when he did. It did not show up until the last day of the conference, and then it was in complete disarray, as if it had been searched thoroughly.

Vigier has been proposing an experiment that would separate the wave from the particle—something that physicists have never believed was possible—using the principles of neutron interferometry. A neutron is a subatomic particle existing in the nucleus of nearly any atom. Vigier believes that a beam (an intense stream) of neutrons can be reduced in strength

So if the wave and the particle are somehow the same thing, then how does the wave suddenly change into the particle? One answer is that objects undergo quantum jumps, like Mexican jumping beans, every time any of them is observed.

Does the Universe Quantum Jump?

Does the universe quantum jump? So far physicists cannot say. The Copenhagen interpretation would say yes. A quantum jump—a discontinuous change from one state to another—must occur every time a measurement takes place. Thus any sequence of measurements would, to Copenhagen enthusiasts, mean that the object looked at quantum jumped from one observation to the other.

They cannot explain in any deterministic sense how such jumps occur. They need the jumps in order to conform to the predictions of quantum physics. They calculate the jumps by calculating the probabilities for the jumps to take place. They assume that the observer is some gigantic memory unit that somehow lies outside the laws of quantum physics.

But for the parallel worlds supporters the story is quite different. Memory exists in each parallel world as the sequence up to the moment. There is no parallel world more real than any other, no matter how freaky the memory sequence. The "buck stops here" sign that was on Harry Truman's desk signifies in the parallel world model the sequence recorded in any world—no world is any better or worse than any other, but all sequences are subject to the laws of statistics.

---

until only one neutron is present. According to quantum theory, even so, that neutron will interfere with itself. The neutron's quantum wave would be divided into two parts, much like a wave passing through two channels, while the neutron itself is posited to move along only one of those channels. Vigier's experiment will test the wave-particle duality hypothesis by separating the neutron from its wave.

There are no quantum jumps in the parallel worlds—just continuous transformations, growths, if you will, of branches of possibilities from a central trunk of initial conditions to a universe of all possible universes.

The parallel universes model has even more surprises in it. If we regard the human mind, for the moment, as a simple recording device—no different from any other recording automaton, like a tape recorder—a new possibility occurs in the parallel worlds model that is certainly not present in the Copenhagen view. It is possible, from a vantage point in the present, to reconstruct a past that is consistent and rational with the present. This may not have been the "real" past, but no one would ever know it. It may also be possible to construct a computer memory that operates according to the parallel worlds view and not the Copenhagen view. Accordingly, a parallel world computer memory would be capable of observing itself even while that memory contained two parallel and quantum interfering streams of data—each stream in a parallel universe. Such a computer would probably use magnetic flux quanta instead of today's chip memory. It would be macroscopic and not atom-sized, however. Consequently, although it would be a large object, it would follow the rules of quantum physics and not Newtonian mechanics.

I am not sure what followers of the original Bohr interpretation would say if such a computer memory were built. Here both the observer—the computer memory—and the observed are one and the same. Using the Copenhagen interpretation, how could one explain the operation of the device? According to its tenets, the observer follows classical physical laws and the observed—quantum physical laws. Here we would clearly be faced with a violation of the Bohr view as espoused by the Copenhagen school.

With the advent of new superconducting devices and the ever miniaturization of computer elements a new class of quantum mechanical automata now appears possible in the near future. We'll explore these questions of mind, memory, computers, and ethics in Part Six. In Part Two we will look at how we can make sense of unity when there is an infinity of universes to deal with.

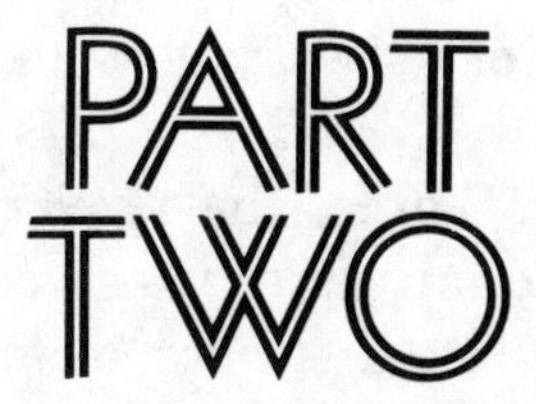

A REAPPRAISAL: WHAT HAPPENED TO THE UNITY IN UNIVERSE?

The Universe—some information to help you live in it.
Population: None

It is known that there are an infinite number of worlds, simply because there is an infinite amount of space for them to be in. However, not every one of them is inhabited. Therefore, there must be a finite number of inhabited worlds. Any finite number divided by infinity is as near to nothing as makes no odds, so the average population of all planets in the Universe can be said to be zero. From this it follows that the population of the whole Universe is also zero, and that any people you may meet from time to time are merely the products of a deranged imagination.

Douglas Adams
The Restaurant at the End of the Universe

If we accept for the moment that any single universe could not occur without the springing into existence of other parallel worlds, other universes, we are faced with a new conceptual problem. Our whole concept of the universe needs revising. Before such considerations as the meaning of quantum physics and all of its weirdness, as we saw in Part One, and Einsteinian relativity and its twists in time and space, as we shall see in Part Three, the universe was the only *one* thing there was. Two universes was a contradiction of terms (an oxymoron). How could there be more than one *everything*?

With the new physics concepts, the universe is no longer the *everything* it once was. It is paradoxically everything and not everything at the same time. How can that be?

The old universe, our home, which we all love and wonder about, taken to be all matter and energy and all of its interactions playing on a stage of existence called space and time, *is* what we meant by *is*—every-*THING*.

But then came the quantum and spoiled all that. Archly, quantum physics requires the existence of that which isn't in order to explain that which is. Furthermore that which isn't affects that which is. So in a sense, that which isn't also *is*.

From the point of view of the new physics, things that aren't play a role in the world of everything that is. The grasp of this requirement of quantum mechanics is probably the greatest stumbling block in the way of understanding just what the new physics is talking about.

In the old physics, the kind mom used to make peanut butter sandwiches with, a thing was a thing. What we predicted about it and what it did, when it did anything, were supposed to be the same. If our theory was right, a thing did what it was predicted to do. If our theory was wrong, then the thing would behave differently than expected and we would trot out our paper pads and find a new theory. In other words, there was a one-to-one correspondence between what a thing did and what we predicted it would do. But that was long ago. Things have changed, the world is more complex, and so are our theories.

In the new physics a thing is represented by all of its possibilities, even those that may be remote. A thing, such as an atomic electron, does not fly through the air with the greatest of ease any longer. Indeed, it doesn't seem to fly anywhere. It just pops up from time to time at various locations that we aren't able to predict with certainty. Or it just sits there occupying no actual space but existing in a ghostlike cloud surrounding the atomic nucleus. To know an atom is to know all of its electrons' possible locations in space. And even that is not enough, for an atomic electron has two guises. When it exists in one guise the other becomes a disguise. It can be a particle occupying a single location at a single time and have an unknowable energy, or it can be stably nested within the atom with a well-defined amount of energy and, as a ghostly cloud called a quantum wave pattern, not have any pointed position at all. Guised as a particle it has no energy. Guised as a wave-cloud it has no position. This is one form of the *wave-particle duality.*

And what's true for the tiny electron is true for all of the objects within the material universe, since all things are made up of electronlike things.

In such a paradoxical universe, it is no wonder that scientists must proceed carefully. To make an analogy, consider a mapmaker's task. Suppose the mapmaker considers the contemporary motorist. For each and every street, a map maker draws a line. To signify a rise in the land, he or she must include contour curves. Finally, when the map is finished, a motorist should be able to use the map to determine how far to drive, when to turn, and where a final location will be.

But suppose the mapmaker considers another type of user, say, a bicyclist or a walker. It makes a difference if the map user decided to motor across the city, walk, or ride a bicycle. The user will need a different map to follow a course and arrive accordingly. The experience of the user, quite different in each case, is accounted for by using a different appropriate map. The user will see an alternate reality depending on how he or she moves through the territory. A bike ride will entail a view of the contours and dirt trails threading the city, a view quite different from that of a car ride. Each map will, accordingly, reflect that change in viewpoint. Thus the user must have an alternate map for each possible use. The bike rider's map is not the same as the car driver's. And yet if we look at all of the maps together, we see an obvious pattern.

In a similar manner, a proper map of the quantum world must take into account all of the ways that a user will engage in investigating the universe. In the new physics, physicists must proceed by taking into account all of the possibilities without prejudice or a Newtonian mechanical law to guide us. But, unlike the map analogy, there is no single trail or road going from here to there. Instead there are an infinite number of trails that we can follow, all starting at one place and time and all ending at another place and time.

There is literally an infinity-to-one correspondence between what we predict and what we observe. We need an infinite number of maps to guide us through the quantum universe. And that would seem most formidable.

But these infinities are within the grasp of human intelligence through the invention of mathematics. Of course, most of us aren't mathematicians, so we remain puzzled and a little fearful of the icy cliffs surrounding the cold quantum universe. Yet, if we bravely scale the heights, carrying our quantum maps with us, and look out at the view provided by the mathematics, we find the view inspiring.

But we need the maps more than ever to appreciate and grasp the view. However, these aren't ordinary maps. These maps are transparent. We can lay one atop the other. And when we do so, we see patterns of sensible knowledge that we can use to make predictions with. Infinity is reduced to a single equation. A new map of the space of our mathematical imagination is conceived containing infinity as a mere point.

But these aren't just mathematical exercises designed to confound our minds. Such concepts as infinity and a space that "contains" all possibilities including other universes, called a *superspace,* appear to be necessary, indeed indispensable, to describe the real world.

Once an ancient philosopher known as William from the town of Occam discovered that people's thinking was fuzzy in regard to logic. He argued that one must use every contrivance to rid us of unnecessary assumptions. He used the analogy of a razor that cuts away all superfluous thinking as it shaved a citizen's face. In his view the universe should have been simpler. William of Occam's razor be damned. The world is complex and we might as well admit it.

New concepts are just what is needed to handle the complexity of our lives. Thus we grow our beards longer in introducing more than one universe. And other universes are just what is needed to understand our precarious position in this universe.

To understand parallel universes we need now to add to our language and to our imagination. In Part Two we examine our present thinking about the universe and how to alter it in order to take into account the new thinking required by the existence of parallel universes. A key insight into the basis for this new physics is self-consistency rather than casuality. In other words, instead of predicting the behavior of physical matter based on what happened in the past, we base physical behavior on the principle that whatever takes place must be consistent with itself. To talk about self-consistency we need to look at how parallel universes change our concepts of infinity and self-reference. We also need to look at *superspace*—the space of spaces and how it provides a new vision of order when parallel universes are included.

Chapter 6

THE BUSINESS OF ISNESS

Complexity, complexity, the word itself confounds us. There is no easy way anymore. Einstein once said, "The most incomprehensible thing about the universe is that it is comprehensible." Well and good, for someone like an Einstein to say. But for the rest of us mortals, it still looks incomprehensible, even to the brightest physicists.

What are we to do with this complexity? One of the hidden axioms of physics is that beneath everything lies simplicity. Whatever secrets lie in store for the discoverers of the universe's laws, those secrets will be simple ones. Oh, if it were only true. But it appears now that we are headed in the direction of more incomprehension, not less.

"Isness" is a complex business, unfortunately. My task in this book is to clarify this view of this magnificent complexity to the best of my imaginative ability. Not only is the universe more complex than we ever thought, it is far more mysterious and magical than we ever believed. Oh, I know I risk the wrath of those who seek simple mechanisms to replace that mystery. But starting out as one of those who so sought, I now see the universe as a gigantic magical mystery tour, far in excess of the Beatles' verses.

And to add to our perplexity we need new concepts like *superspace* and infinity—that mysterious concept that, like an eager tour guide, leads us ever onward, never letting us stop and rest and peer at the sights. Infinity is always one more than now.

Infinity, Infinity Everywhere and Not a Spot to Think

Many people have a great deal of difficulty dealing with their ordinary world. Each of us has chores to do each day, a job to go to, a bank account to balance, and a seeming infinity of never-ending bills to pay. This is probably the closest most of us come to the idea of infinity—something that goes on and on and never ends.

Some scientists are also bothered by infinities, which seem to crop up at embarrassing places in our theories of the universe. A black hole, as we shall see in Part Three, has an infinity at its very center. There, nearly everything calculable or measurable grows to unmanageable size—like our credit card bills at the month's end.

In the description of an electrically charged particle, such as an electron, we find infinities lurking. They arise in the mathematical description of electrically charged particles that is called *quantum electrodynamics*. What emerges in that description is a story in which the electron seems to be aware of its own existence. It interacts with itself like any bad little self-abusive boy behind locked doors. When it does this it generates infinities—an infinite amount of energy, for example. But when mother-physicist comes home and opens the atomic door and observes the electron, the little angel is peacefully obeying the rules of the universe.

Physicists have discovered tricks for getting rid of these mathematically generated infinities. But, nevertheless, they are there to plague us and they are needed whenever we calculate something involving the quantum electrodynamics of electrons.

Now this bothers many scientists. For some, perhaps perverse, reason, it doesn't bother me at all. I like infinities. I believe that infinity is just another name for mother nature. Nature provides infinite possibilities all of the time. But because we have suffered through this world of wars and woes, we sometimes fail to get this. We see the world as a little stingy at times.

But we shouldn't give up too much hope. If quantum physics and relativity are correct, then infinities are real, just as real as mom's apple pie.

Now why do we need a conceptual grasp of infinity in this book? Because, according to the parallel universe idea, there must be an infinity of universe-possibilities just to have this one reality. I believe that the infinity of possibilities predicted to arise in quantum physics is the same infinity as the number of universe-possibilities predicted to arise in relativistic physics when, at the beginning of time,* the universe, our home, and all of its sisters and brothers were created. As modest and troublesome as we often are, we too are nevertheless creatures of infinity.

To grasp how infinity and parallel universes are related, I'll take you through two visual metaphors in this chapter and one logical one in the next chapter, each exhibiting an aspect of infinity relevant to our understanding of parallel universes. I'll try to point out, in these examples, how nature generates infinity as naturally as she makes a tree.

*See Part Four for a description of how parallel universes were generated at the beginning of time.

The First Infinity: A Straight Line

The first infinity is not too difficult to picture. Imagine a line one inch long drawn on a sheet of paper. Suppose that this line represents the whole universe. Now that's an abstraction—something physicists create whenever anything is complex. In other words, each and every point on the line corresponds to a single particle in the universe.

Now scientists have estimated the number of particles in the universe. Based on some astronomical observations, the Milky Way galaxy—the galaxy in which our solar system exists—has around 10^{11} stars.[1] Stars, like little lost children seeking shelter on a cold night, tend to cluster, via gravitationally induced starlight, into galaxies.

Galaxies themselves tend to cluster in groups called supergalaxies. We are not quite sure why they do so. No doubt gravity has something to do with it, but it isn't that easy to figure out. Each supergalaxy has a mass of 10^{14} times the mass of a single star such as our sun.

The mass of a single star can be estimated. And from that, knowing the mass of a typical solar particle, such as a proton (a nucleus of the hydrogen atom), we can determine the number of protons inside our sun—which, to be sure, is a typical middle-aged star. Based on these observations and a few calculations, physicists have estimated the number of particles in the universe to be 10^{80}, plus or minus a few.

Now this number is quite vast. It is difficult to imagine a situation to which this number can be compared. But yet, as difficult as this is, the number of points on a simple line, one inch long, on a sheet of paper, far exceeds this huge number.

In fact the total number of points on the line is infinite. To see why, one must imagine a procedure for counting and distinguishing one point from another. This would be easy if a point could be grasped. It would be duck soup if a point had a finite size or mass or both.

But a point occupies no space at all. It has no mass to speak of. Just lying there on a line, all one can do is assign it a very specific number such as .5 or 2/3. This number could be, for example, the distance from one end of the line to the point in question.

But why are there an infinite number? you might ask. To see this, imagine looking at two points that are very close together. As long as they don't touch, which means that two different numbers can be assigned to the two points, and one of them is farther away from an end point of the line than the other, then there is always a third point that can be found in between.

In mathematics we say that between any two numbers there can always be found a third number greater than the smaller of the two and less than the larger. Now consider the two closest points among the three. Again, if they are separate from each other, another point can be found in between once again. This procedure of in-betweenness can be extended, and each time the two closest points will always be found with another point in between. In fact, the idea that a point takes up no space itself is the same idea that two points represented by different numbers cannot be the same point. Since you can always find a point between any two points, there is an infinity of points on the line.

Infinity after all is a concept. And it is conceptually graspable as a procedure rather than as a thing in itself. Infinity is always one more than now.

Infinity Number 2—It's All Done with Mirrors

Now take another example. Picture yourself standing in front of a mirror. Parallel to the mirror and on an opposite wall is another mirror. Now suppose that the mirrors are perfectly reflecting. This means that each and every photon of light that strikes a mirror is always reflected. (A real mirror won't do this. The best real mirror may reflect ninety percent of the light striking it.)

Of course, you can see yourself in one of the mirrors. But so can the other mirror "see" the mirror in which you view yourself. It also "sees" the reflection of yourself coming from the first mirror. Each photon carrying your image reflecting from mirror one reflects from mirror two. But mirror one also picks up light coming from mirror two. So it "sees" the image of the image coming first from itself and then from mirror two back to itself again. And mirror two "sees" this image—the image of the image of the image.

And off we go to the infinity races. Always one more than now. If you can manage to move one of the mirrors slightly so that it isn't exactly parallel to the other, you can witness the infinity effect.

It is quite weird to see yourself reflected an infinite number of times—a hall of mirrors. Each image is an exact duplicate of the first image, but each image grows smaller and smaller as it reflects the image from the distant image coming from the other mirror.

An insight here is that infinity can be generated from two, provided the two provide a feedback loop—a self-consistent reference. Each mirror provides a reference for the other. But since each carries the same image, it is a self-reference—in other words totally consistent with itself.

Chapter

7

HOW THINGS THAT ARE DEPEND ON THINGS THAT AREN'T AND HOW WE AREN'T NORMALLY AWARE THAT ANYTHING IS ANY DIFFERENT

Like the parallel mirror example, all self-reference paradoxes are generators of infinity. And the self-reference paradox helps us to understand how it is that we are normally not aware of other universes even if they exist side-by-side with our own. Let me explain this with an illustration.

Imagine a small card. Written on one side of it is the sentence:

A. THE SENTENCE ON THE OTHER SIDE OF THIS CARD IS TRUE.

Now turn the card over. On the obverse side it reads:

B. THE SENTENCE ON THE OTHER SIDE OF THIS CARD IS FALSE.

Infinity occurs when you place yourself in the loop and put meaning into the sentences. You read each sentence and then logically construct its meaning. This sets up a dynamic movement in which the truth or falsity of a sentence oscillates between "yes" (meaning true) and "no" (meaning false).

The A sentence, if it is true, instantly makes the B sentence true. But the B sentence says that the A sentence is false. Suddenly there is a quantum shift in our perception. We instantly realize that the A sentence is now lying about the B sentence. Since the A sentence says the B sentence is true, and the A sentence is now false, the B sentence must therefore be false. This means that the B sentence now lies about the A sentence by saying that it is false. Therefore, the A sentence quantum leaps from being false to being true. Here the pattern is not self-consistent. Consequently it only appears as an amusing puzzle. However, the whole pattern repeats, on and on into infinity. Infinity is always one more than now.

Reflections in a Parallel Universe

These examples of infinity are reflected in the parallel universe theory and are important in understanding how parallel worlds are reflected in our own. The last example was not self-consistent. The game of physics is to invent principles that *are* self-consistent. It appears that the rules governing parallel universes are self-consistent. (To avoid any confusion for now, think of the word "universe" and the word "possibility" as meaning the same thing.) They illustrate three principles of the new parallel universe vision of reality. Put briefly, reality in the here-and-now depends on all of the there-and-thens.

All infinity of them.

These principles are

1. In-betweenness: There is always an in-between, and therefore infinity is real.

2. Reflection: Parallel realities are infinite reflections of any one reality.
3. Self-reference: A self-consistent referral generates, from parallel universes, a single universe.

In-Betweenness or Room for One More

The first principle, in-betweenness, means that from any two possibilities there is always a third possibility and the process is never-ending. In the quantum world view, these possibilities continually arise as objects interact with each other and as observers of objects continually interact with them.

Let's look at an example. Suppose you are measuring a location for an atomic electron. To do so, you, the observer, must interact with the electron—you must perform a complex set of operations which enable you to determine where that tiny particle is located.

In the old physics, this was nothing more than a technical problem. The electron had a single location—it had to be somewhere—and your job was just to find it. But in the new quantum physics, the electron has no definite position or, in other words, it has all possible positions simultaneously. But, in a self-consistent manner, each possibility must exist in a separate universe.

When you find it, each of those possibilities manifests simultaneously. That's right, all infinity of them. In each possibility or, if you grant me the license, universe, there will appear a single electron and a single viewpoint—your own.

Your world before the electron measurement was one universe. The electron's world was many universes. After you observe it, your world after the observation is many universes. You become correlated with the electron in all of its universes. Your mind is split into myriad minds, each intelligently recording its observation.

Yet, surprisingly, nothing strange is going on. No splitting is felt. In fact, all the splitting occurs only in your own brain. The net result: you are aware of all of the measurements at once. Most of these results will differ by hardly a hair. Your normal waking consciousness will fail to distinguish one result from the other. The results will form a pool, a superposition, a gathering of viewpoints, and you will be present in each component and not notice that any difference of viewpoint exists.

In between any two possible results, any two universes, there will always be a third result present. The sense you make of any of the results depends on that fact. The closeness of the results to each other produces a resonance. Just as a soprano's high note causes a wine glass to break, two or more universes playing together resonate. Something solid arises from the two together. If the results were not similar and self-consistent, the data would appear scattered and random. The sensibility, the comprehensibility, of the experience depends on the closeness of the results to each other. The greater the similarity of the observations, the more solid the object appears. Reality is nothing more than a lot of agreement.

Reflection

The second principle, reflection, I think of visually. I imagine parallel universes as parallel mirror reflections of this universe. In each mirror there is a you and there is an object. I and it, it and I. And in each universe there is a question: Why is this universe different from all other universes?

Like the mirror situation where the mirrors reflect an image of me, the parallel universes are somehow duplicating something real—a real universe. But what makes one universe any more real than the others?

Suppose we made a checklist of all the universes—all infinity of them. In most we would find duplicates of ourselves hardly differing from each other. There would be differences, here and there. In one you might be wearing a solid yellow tie, while in the other your tie could be yellow-brown.

But like transparent slides viewed simultaneously, these universes would actually superimpose together and appear as one. This superposition contains our minds in each one, and since our minds do not distinguish between the universes—that is, our normally attuned minds don't—we fail to notice that each and every one of us is an infinite family. In the case of the electron, we would know, for example, that the electron was in a particular atom, but have no idea where it actually was located within that atom.

Taken together, all of the infinite possibilities for the electron, all of the infinite universes in which the electron occupies a single point and no other, the electron is nowhere for sure. But it is in a stable energy state. The atom is normal, but the electron is schizophrenic, metaphorically speaking. Thus we can view all of the infinite number of position-possible-universes as composing one single energy universe.

The gathering of universes into a normal universe accounts for our presence. Without such coalescing, there is no stability possible. The universes we inhabit contain something apparently unique. Yet there seems to be a hidden principle at work. There is a matching of gravitational energy with the expanding energy of the big bang, for example. Some scientists argue that there does, indeed, exist a guiding principle which fine-tunes the cosmos to such incredible accuracy. This is called the *anthropic principle.*[2] It says that from an infinite number of possibilities nature could have selected to make a universe, it selected this one so that we could be created. If we add to this the parallel universe idea, then nature plays it safe. She creates all universes, even those where consciousness doesn't exist.

What about the others? They mirror our own but with slightly different values. Yet when all the differences are taken into account (we say in the superspace of all universes), there appears to be an order arising. This order causes many universes to appear with just the right values to make our universe real.

And that reality is just what is needed to make it possible for human consciousness and indeed all life to manifest. Thus, in some strange way, the infinity of universes appearing is just nature's way of having some large enough number to choose from in order to assure that conscious beings make their way onto the cosmic stage. For what purpose, we can only create.

Chapter 8

SELF-REFERENCE: ISNESS AND AIN'TNESS

The third principle, self-reference, helps us to grasp why we are normally aware of only one of the parallel universes even though quantum physics predicts that an infinite number of them exist side-by-side. It turns out that the reality we observe depends on us—on how we choose to refer to it. Ultimately, reality is a question of self-reference—what we choose to be outside of ourselves and what we choose to be ourselves.

The reason for this lies in the way in which quantum physics predicts reality—how any object of our everyday experience—arises. According to the quantum rules, we cannot ever know and experience simultaneously all that is in principle knowable.

Why we face this perverse state of knowledge and existence is not known. Of course, we know how to handle the rules governing the game of knowledge and existence. Those are the quantum rules. But we cannot fathom why they should be so strange. One thing is clear, though: self plays a role in what is seen to be not-self.

The quantum rule governing this is found in Heisenberg's *principle of uncertainty.* This principle governs the game of knowledge and existence. In its major application, although either the location of an object or its path as it moves through space and time is in principle knowable, both cannot be observed simultaneously. The actions taken by an observer to determine the object's location necessarily cause the object to *split* and journey along many separate paths simultaneously. The actions taken by the same observer to determine the path followed by an object render its position unknowable—or, in other words, the object with a well-defined path also *splits* and has many positions simultaneously.

The old adage that a thing cannot be in two places at the same time is only half-true. It is more accurate to say that it cannot be observed in two positions at the same time when an observation of its position is carried out. Actually, it can be observed to be in an infinite number of positions at the same time. All one needs to do is observe the object's path, and its positions are all there—no single position is discernible.

Perhaps it is useful to look at an analogous example of the path-position complementarity. Suppose that you consider a camera. You can determine the location of an object, say a high-speed bullet, by taking its picture. Of course, if the object is moving fast, your picture will be blurred unless you use a very rapid shutter speed. The more you attempt to "pin down" the location of the object, the faster your shutter speed must be. But you pay a price for accuracy of location. The bullet doesn't appear to be moving in the photograph. In fact, you can't be sure of its path—its direction through space—at all.

On the other hand, by allowing a slow shutter speed, you do capture a blurred line showing the trajectory of the bullet. You do photograph the path, but you lose the object's position by doing so. Gaining the path you lose the position, and vice versa.

Similarly, other attributes of quantum objects are incapable of being observed simultaneously. However, it is important to remember that what isn't observed (what doesn't manifest) in this universe exists side-by-side in other parallel universes. Thus what exists depends on what doesn't exist. Or it depends on what we choose to observe and refer to.

It's a Complementary World After All

When two or more attributes of an object are incapable of being observed simultaneously, the attributes are said to be complementary to each other.

There are many examples of complementarity. The directions of spin of a subatomic particle are complementary to each other. An object spinning with its axis pointing horizontally must also be spinning with its axis pointing up and down at the same time in separate parallel universes. Similarly an object observed with its axis pointing downward must also exist with its spin axis pointing to the right and to the left simultaneously in other universes.

The key insight here is that each observation brings to light one of a set of possible values of an observable attribute (simply called an observable in quantum physics) and at the same time causes the other complementary observable to assume all of its possible values simultaneously. But not in any single universe.

These other values are not seen when the original observable is seen. The others are, however, observable in parallel universes.

An Example: A Coin in Four Universes

To explain this it will help to use a simple fictitious example. I'll use this example from time to time throughout the book, each time illustrating a different aspect of parallel universes. Imagine to begin with that there are only four parallel universes. To make things as simple as I dare, imagine that all that exists in each universe is an observer and a tiny quantum coin.

This coin has only two observables, its position and its color. The coin can be observed with its position of heads or tails up. Or the coin's color can be seen to be red or green. And these observables, color and position, are complementary to each other. Thus if the coin has heads up, its color will be uncertain. That means that its color is red in one parallel universe and green in a second parallel universe if it is found with heads or tails showing in this universe.

Let me clarify this somewhat. Now this atom-sized coin is *not* an ordinary coin. I am sure that you could see both the color of a copper penny and whether or not it had heads face up. Both the color and the position of the coin exist in a single Newtonian universe. An ordinary coin follows Newton's rules of decorum and needs no parallel worlds to complicate it.

But the quantum coin is very, very tiny. It follows the rules of the quantum world and, to the nonscientist reader, what follows may seem to be a little like a magic trick. I assure you that I am not trying to pull the wool over your eyes! Think of the coin's position and its color in the same way that you might think of an atom's position and its momentum. Any attempt to observe the coin's color necessarily disrupts any attempt to observe its position and vice versa. This means that according to the parallel universes interpretation, both color and position cannot exist in a single universe.

In the parallel universe model we dispatch with all of this

uncertainty by putting all possibilities in separate but equal parallel universes. And we also put the observer of the coin in each of those universes observing the chosen attribute. This wasn't necessary in the Bohr Copenhagen way of seeing things. There the observer always existed in a classical Newtonian universe—this one! When the observer saw the object, the object instantly took on a value associated with what was seen. This is no longer the case in the parallel universes way of seeing.

Now, this is a little bit tricky. There is, as far as the observer and the coin are concerned, only one observer and one coin. Each universe represents a possible relationship between the observer and the coin.

There are four possible relationships:

one: the coin has heads up
two: the coin has tails up
three: the coin glows red
four: the coin glows green

Thus each universal possibility appears to exist as if the other weren't really present. It is here that the requirement of self-reference enters the theory. Here we see why quantum physics demands the other universe's presence.

First, the sides of the coin must be *seen* to be believed. And to do this, the observer must use light. However, to see a detail like a heads or a tails on a tiny quantum coin the observer must use high-frequency, short-wavelength light. Such light is beyond the normal color vision. And that light has a momentous impact when it strikes the coin. Sure you can see which side is up, but, as in using the high-speed camera in the path-position problem above, you cannot tell what color the coin has.

But this tiny coin also has a color. If you shine low-frequency light waves on it, the coin will glow red or green. However, using low-frequency light, as in using a camera with a slow shutter speed, you cannot tell which side of the coin is facing you. You can only see a color. The longer wavelengths of the visible light spectrum will clearly let us see the coin's glow but won't enable us to tell which side of the coin is up, because the long wavelengths are too long to make out the finely spaced details of heads or tails.

Now let's suppose that our observer decides to measure the position of the coin and find out which side is up.

In universe one, the observer observes the position of the coin to be heads. But according to the parallel worlds rule, in the second universe the observer sees the coin with tails showing.

The observer in the first universe, observer one, attempts to make sense of his observations. He objectifies his experience by including himself in the process of determining the position of that coin. He puts himself in the state of consciousness: That is a coin's position distinct from me.

Similarly, observer two in universe two, in determining that the coin has tails showing, observes that it does not show heads. According to ordinary logic, the coin cannot have both heads and tails showing simultaneously if it is being observed in a position state.

Each observer self-referentially observes that he is the only observer and that the other result for the same observation doesn't exist.

But in parallel color universes, three and four, the same observer decided to observe the color of the coin. In universe three, the observer saw the coin glow red and in universe four he saw the coin glow green.

My point here is that once an observer takes an action, *all possible paths* for that action emerge. Thus four distinct universes, one and two (connected with the position observation), and three and four (connected with the color observation), exist, all simultaneously and each separate from the others.

But why do we need all of these universes? In the quantum

view, we need them because of the connection that exists between one set of observations and the other complementary set. We see that the color of the coin and its position are complementary to each other. When color is observed and determined, so that an observer in each of the universes three and four sees and believes that he sees a unique color, a complementary event is happening in universes one and two. In each of those universes the observer sees and believes he sees a unique position of the coin. Both positions of the coin and the observer of those positions must be present simultaneously.

Clearly it is impossible in any single universe for an object to exhibit two or more contrafactual attributes at the same time. A coin showing heads and tails simultaneously would not be a coin showing one side up and the other down. What could it be? The answer is a coin showing a color. Thus any universe is the confluence of agreements concerning what is logically consistent and nothing more. That which is logically inconsistent appears as another complementary facet, which in itself is logically consistent.

Thus if you see a color of a coin, you are simultaneously seeing heads and tails. Only you don't think of the simultaneously observed contrafactual positions of the coin as an impossibility. You experience them as a real attribute—the coin's color. This is analogous to the situation concerning the stability of the atom discussed earlier. For the atom to be stable—exist with a defined energy—the electron must exist as a cloud. Thus all electron positions are simultaneously observable in order to have one energetically stable atom.

Thus, when any of the physical attributes of the coin is observed, say, the position of the coin, all four universes spring into existence. However, universe one is really an overlap of universes three and four. Similarly, universe three is an overlap of universes one and two. So actually, even though there are four universes present, never is more than one universe actually perceived. The complementary universes are just as real for the observers in them as they are in any universe. They are all connected.

A single universe's existence depends on the existence of its sister parallel universes. It's a weird world, albeit, but if we believe quantum physics, it is necessary and sufficient that it be this way.

Chapter 9

LOOK UP IN THE AIR— IT'S SUPERSPACE!

If there are really other universes, then just where are they? How far are they from midtown, and can I get a good corned beef sandwich (with mustard, no mayo, please) in one of them? If they are so far away as to make no difference, why worry about them? Indeed, who can believe that they are there anyway?

I believe that they are *here,* now, inhabiting the same space as we live in. Space is far weirder than we ever imagined. Just as a subatomic particle, like an electron, can exist in more than one place at the same time, and yet never be observed except at a single place at a single time, space itself can exist in more than one *space* at the same time, and not be seen any differently than we presently see it.

In other words, these spaces, these whole universes, overlap, as if they were nested together like Chinese boxes one inside of the other. The only difference is that the boxes are all the same size!

An object in any space exists in all of the spaces at the same time in the same way that an electron in an atom exists at an infinite number of points at the same time but only occupies a single point any time it is observed. Objects in each space pass through the other spaces like ghosts in the night. Yet these objects are totally and consistently solid in their own separate spaces.

Just to imagine such a thing as this is difficult, if it is possible at all. One way is to bring in an idea of a space that contains all possible spaces. It is called *superspace.* There is always room for another universe, another whole space, in superspace. In it, objects take on individual but separate realities. And all of those realities move like ripples in the wind of probability.

Superspace! Broadway Theater Style

Superspace is an imaginary mathematical structure used to envision situations in which there exist more than three dimensions. Physicists attempting to put relativity and quantum physics together in one package came up with the idea of *superspace* just in the past thirty years or so.

Superspace contains points just as ordinary space does. But each point in superspace marks the locations of every object in a whole universe. That is, each point in superspace is a scale model of a whole and distinct universe. Of course, such a space is a bit weird, and anyone would have difficulty imagining such a thing, but I'm going to try to paint a picture of it for you as best I can.

To begin with, it has an infinite number of dimensions, and infinity, as we have seen, is no easy thing to grapple with. However, superspace can be visualized if we first imagine how you might construct a layout map showing where Broadway theatergoers would sit in a specific row at a sell-out.

First you draw parallel lines on a blank sheet of paper representing the rows, and you leave gaps marking the aisles. You notice where one spectator is located with respect to one end of a particular row by placing a dot on the drawing. The dot shows his location to scale in the model by being marked off an appropriate distance from the end point of the line which denotes the aisle.

For example, if playgoer A is five chairs from the aisle, you might scale the model using one inch per chair and pencil in the dot five inches from the starting point of the line. If another playgoer, B, sits in the same row, say, seven seats from the aisle, you would repeat the procedure only using the playgoer's distance to mark her spot.

Now that's just fine for two persons in a row at a theater. One needs only a one-dimensional line to describe the location of each person. Person A is five chairs from the aisle, and person B is seven chairs from the same aisle.

You could then imagine repeating this procedure for each person in the row. Your model of the theater row would be quite simple. One dot on the line for each spectator. Ten spectators, ten dots, for example. Compared to a universe, a row full of people is a cinch. The only problem is that you need a lot of dots.

There is another way to model the people in the row using only a single dot. This is the superspace way of doing it. You up the number of dimensions in the model by one more dimension, imagining the model of the row to now be a two-dimensional square instead of a one-dimensional line.

Why would anyone go to this trouble? you might wonder. A one-dimensional line is much easier to deal with than a two-dimensional square. However, there is an advantage to upping the dimension of the model. A single point inside the two-dimensional square actually tells us where two people are sitting.

Suppose you wanted to locate the two theatergoers, A and B, in the row just using a single point in the model. If A is five seats and B is seven seats from the aisle, you could pencil in a dot that was five inches from one edge of the square and seven inches from an adjacent edge.

The single point in a higher dimensional space actually appears as a greater number of points in a lower dimensional space. Thus a single point within the interior of the square tells us where two people are seated.

Similarly if you modeled the row by a cube, you could describe, with a single point, the location of three people in the theater row. The distance from the floor of the cube to the point would tell you where person A was sitting. The distance from the north wall to the point would tell you where B was sitting, and the distance from the west wall would tell you where C was sitting.

Each time you add a person to the row, you up the dimension of the model. With a full row of ten people, you use a ten-dimensional space, and with a single dot you mark the locations of all ten theatergoers.

If you needed to note which row as well as which seat, you would need a twenty-dimensional space to mark the ten theatergoers. And if you needed to know on which floor of a high-rise the theater was located, you would need thirty dimensions, but still just one point. Although the space is increasing in dimensionality, a single point tells it all.

Since there are more than 10^{80} particles in any single universe, the model of superspace must contain more than 3×10^{80} dimensions (the factor of three to take into account the three dimensions of ordinary space).

But if we take all of these dimensions into account in the same way that we did for the theatergoers, a universe with all of its particles in superspace is just a single dot in the space.

Uncertainty in Superspace

However, according to the uncertainty principle of quantum physics, you can't ever know such a thing as the exact location of any one of the objects in a universe. Such knowledge would be at the expense of using extremely high-energy devices* that would be needed to pin down each object's position. Not only would this be technically impossible to do, but having the exact location of every object in the universe would cause extreme violence.

Heisenberg's principle tells us that information about the location of each particle would introduce complete uncertainty in the momenta of the particles. Remember that momentum means the product of mass and velocity. An object with a large momentum can cause damage when it smashes into other objects. The uncertainty in momenta of particles means some of those particles would have explosive energies. Even a small fraction of 10^{80} particles with explosive energies would annihilate the whole universe.

To render the universe benign, to keep energy conserved and not wandering all over God's creation, the objects in the universe we inhabit can't have exact positions. Now the problem is, just what does that mean? No one knows for sure. In this book I've taken this to mean that a single stable-energy universe must consist of an infinite number of parallel particle-located universes. In each of those parallel universes each object exists with a fleeting position, but not with a defined momentum or energy.

Each of these universes fluctuates and each is highly unstable. Objects seem to vanish and appear much like amateur photographers' flashing bulbs at a crowded nighttime football stadium. Yet, put an infinite number of these unstable universes in the same space, and voila!—you have a stable universe.

*You need high-energy devices to locate subatomic particles because of the fine accuracy required. To locate an object within a range of a few billionths of a centimeter requires that you use photons of light that have wavelengths at least that small. To locate a subnuclear object you would need gamma rays which have much smaller wavelengths. The shorter the wavelength required, the more the energy needed. The energy of a photon is inversely proportional to its wavelength.

This is where the superspace model becomes useful. We need the model because a cloud of universes actually exists. A cloud of positional universes is what we live in. We need all the points in the cloud to have any stable universe, in the same way that we need a single electron to exist as a cloud in an atom in order that the atom have a stable energy.

What If Newton Was Right? Silly Superspace

If classical physics actually worked in the universe, superspace would be silly. All one would see in it is a single point moving around as all the particles in the universe changed positions. It would be just a grander version of musical chairs.

Since quantum physics is necessary to describe a universe, a single point in a superspace model, aside from its not telling us more than we already know, would not adequately describe what we see as reality. To attempt to model reality using our everyday three-dimensional space would leave us in the trees lacking a vision of the forest. Ordinary space lacks space. Even in superspace, you need a cloud of points, each point representing a parallel universe. Without the cloud, no single universe in reality would remain stable.

Quantum physics makes things more complex than classical physics because it deals with representations—clouds, waves, distributions—of particles and not particles themselves. However, it is not a hopeless or fruitless task. It turns out to be simpler to deal with a cloud or wave than a bunch of particles, if you use superspace.

Some physicists, however, think that parallel universes and a superspace for them are not necessary at all. They think that quantum physics is nothing more than an extension of classical physics in some subtle way. They would point out that all superspace is needed for is to describe the statistical variations due to our ignorance concerning the locations of all the particles. As they see it, particles do occupy space, time, and matter all at once. They believe in the reality of the physical world, and we can't really blame them! After all, the world seems pretty solid to you and me.

For example, physicist Leslie E. Ballentine, from Simon Fraser University in British Columbia, believes that the idea of infinitely multiplying, noninteracting worlds should be taken somewhat less seriously than the Ptolemaic theory of epicycles.[3]

Ancient Greek philosopher Ptolemy believed that the earth was at the center of the universe. He thought that all the celestial bodies moved in perfect circles, with the earth at the center of all this perfection. Circling around the earth were the planets, the moon, and the sun. Because these *circles* were observed to be markedly different from actual circles, he proposed that the planets orbited in such a manner that they made circular loop-the-loops, with the loop centers following perfect circles around the earth. In this way the heavens maintained their perfection.

These loops were called epicycles. With no other way to explain the apparent discrepancies between the desired perfected circles of Ptolemy and the observed actual motions of the planets, the epicycles were accepted, but no one really liked them very much. It was simply that they appeared to be a required complexity to maintain the Ptolemaic basis of all celestial motion. With the later discoveries of Copernicus, Kepler, and Newton the epicycles were not needed and, as it were, were "chopped away" with the ever powerful razor of William of Occam.

Physicist Ballentine thinks that parallel universes are as unnecessary as epicycles. To him, one universe is quite enough even though it is complicated by quantum rules. Ballentine believes that quantum theory cannot be used to describe a single system. It, instead, describes a fictitious ensemble of possibilities. Thus quantum theory is only a statistical theory in which the objects of statistics are not in themselves real. The real world is there but the theory is incapable of describing it.

Probably the foremost advocate of parallel universe theory is physicist Bryce DeWitt of the University of North Carolina. In a now famous article, first published in *Physics Today,*[4] he describes "quantum mechanics and reality," from the parallel universes point of view. This was one of the first popularizations of the theory—even though it was written for physicists. DeWitt believes that the solution to the dilemma of indeterminism is a universe in which all possible outcomes of an experiment actually occur. In a later paper[5] written for *Physics Today,* in 1971, he took on a number of dissenters, including Ballentine, providing counter-counterarguments. DeWitt added that in his view the number of possible universes need not be infinite—one hundred powers of ten (10^{100}) would suffice.

He pointed out that Ballentine's example of comparing parallel universes with Ptolemaic epicycles was curious. Everett, the inventor of the concept of parallel worlds, in his Ph.D. thesis compared the sensory testimony of those who claim the splitting idea to be absurd with that of the anti-Copernicans in the time of Galileo, who did not feel the earth move. In other words, Everett used the same argument as Ballentine to prove that parallel universes could exist in spite of their apparent insensibility.

In Ballentine's view, since quantum physics was just another kind of statistical physics, it was just another branch of classical physics. In such a classical physics, even though we are able to practically measure only such things as temperature and pressure, the fact that these quantities are averages, taken over many, many particles interacting with themselves and their containers, makes it possible to determine, in principle, the exact locations and momenta of those particles. Thus Ballentine objects to the parallel worlds view that these hidden quantities are unattainable.

DeWitt points out that quantum physics and statistical physics are quite different *in principle* when it comes to what is knowable or attainable. Where Ballentine believed that the many branches predicted by the parallel worlds theory were fictitious composites which described possibilities and only possibilities for a single object or universe, no single object was ever described by a single branch. Where is the real world? asks Ballentine. It is here, somewhere, but we don't know where. The branches, however, are not real if one thinks of each branch as occurring simultaneously. Only one branch is real.

In DeWitt's view, each branch was more than a possibility; it was a reality. In the parallel worlds scene a branch contained both an object or universe and an observer of that object or universe. In each branch, there was a fundamental unattainability of information about the other branches. As DeWitt put it in describing subatomic electrons in parallel universes,

> *Although these electrons are described by the same* [mathematical formalism indicating only a single electron], *they actually inhabit different, not the same worlds, and I prefer to think of them as different. So, when I'm on an aircraft about to crash, I am going to worry. It's me I'm concerned about, not those other guys! In a more relaxed mood, of course, I am quite prepared to take my other selves seriously even if I can never know what they are doing.*

According to Ballentine, because we humans can't ever keep track of all of the particles in the universe, we keep statistical averages, notebooks of data stating where things are likely to be. It is not that the objects do not possess positions; we simply are ignorant of them. As the particles change positions in the only universe that ever was, a cloud of points would be needed to envision the possible movement of the real particles. This is the way that one uses thermodynamics or statistical mechanics of gases to describe a wind or a real cloud of water vapor.

Consequently, DeWitt and Everett believe that quantum physics is completely adequate to describe the real world. Ballentine and others do not feel that this is the case.

A Pattern of Order in Superspace: Consciousness Emerging

If we were to leave Occam's razor in its sheath and explore the superspace of these imaginary unconnected parallel universes—called statistical ensembles by physicists and predicted by statistical classical physics—each point in superspace would move in entirely random ways. It would be like watching a modern video showing several images all flashing together, but making no sense.

However, when we picture a universe and its infinite number of parallel universes in superspace using quantum physics, as it is literally described by the mathematics, a pattern of order emerges. This order is not contained in the motion of any single point in it. In fact, points don't move continuously in superspace, according to quantum physics. They simply fade in and out like light-bulb signs in Times Square. While a single fading in and out is random, from this fading in and out a universal light show of orderly movement is perceived.

It is this pattern that we humble humans are capable of understanding by using superspace mathematics.

However, like Hansel and Gretel lost in the forest, we can quickly lose the thread gained by just using the old comfortable three space dimensions and one time dimension of the everyday world. Things might seem beautiful and so complex that we begin to wonder how any common sense of the universe can be gained at all.

If parallel universes are the only way to bring relativity and quantum physics together successfully—in other words, the only way to have a view that there is a "sane" and objective single universe, why should the universe be so bizarre?

I would like to suggest an answer. This is the only way that consciousness, awareness of the illusion, is possible. Let me explain this somewhat mysterious point.

I believe that consciousness has no place in a classical physical universe. It has no role in a universe that runs like a machine. If there were consciousness in such a world, it would indeed be a by-product of materiality—an emergent phenomenon having no more significance than a rock floating in space—dead to everything.

Quantum physics appears to be telling us that what we choose to observe alters, and even creates, what we observe. Thus in a quantum world view, we have choice—something I see as synonymous with consciousness. In other words, to have consciousness there must be choice.

But how can choice manifest? There must be mind. In other words, it is self-consistent to have choice if there is mind, and choice then exists in the mind. Mind, I believe, exists as fleeting energy in parallel universes. The universe we perceive consists of the overlap of these fleeting flashes of energy. The patterns create mind as surely as they create matter. Both the existence of matter and the perception of it are the same thing.

Thus it is that the mind of any sentient being that is capable of perceiving a reality is capable of reaching into parallel universes and performing the task of choosing that reality.

There is a subtle trap in this. Once a choice is made, the chooser, like Br'er Rabbit and the Tar Baby, is caught by the choice. The chooser must enter, become one with, the universe chosen. This thought bothers me. It gnaws at me. If this is true, do we really have any choice at all? If every time I choose, all of my parallel *me*s are also choosing, then is there really any choice? If any choice means all of them, then perhaps choice is just another illusion. There really is no free choice, because all of the results of any choice always manifest. We may be machines after all, but just existing in more dimensions than we are aware of in any single universe. I ask the reader to consider the possibility.

Chapter 10

A MOUSE, A COIN, AND A QUANTUM CONSPIRACY

When Einstein considered the usual or Copenhagen interpretation of quantum physics, he was, as you may know, critical of it. He was mainly concerned with the drastic changes of an object's physical properties brought about by the simple acts of observation—something that is now referred to as the *collapse of the wave function.* He put his feelings quite colorfully, stating that he could not believe that a mouse could bring about drastic changes in the universe simply by looking at it. Einstein felt that quantum physics failed as a complete theory because it did not take into account the way an observer gained information.

Philip Pearle, a physics professor at Hamilton College in New York, agreeing with Einstein, also felt that quantum physics failed as a complete theory.[6] He followed one of Einstein's ideas that quantum theory describes only an ensemble of possible results. Thus the theory is only a statistical one. To Pearle, like William of Occam, the notion that the world suddenly collapses into one state only when a measurement is performed or continues to expand into infinitely many real worlds is too artificial and uneconomical.

Mendel Sachs, a physics professor at the State University of New York in Buffalo, also felt that an alternative theory was necessary.[7] He provided an interpretation different from the parallel worlds one. Sachs wanted to add a nonlinear mechanism to quantum physics. Its solutions do not fit into superspace. His theory also predicted results that are not predicted by quantum physics. His approach was based on a correspondence between the results of an observation consisting of many observations and the result obtained in a single observation. Just as Einstein's theory approaches Newton's theory of gravity when distances become small when compared with measurements over astronomical dimensions, Sach's theory when completed would encompass quantum theory.

In reply to Pearle and Sachs, Bryce DeWitt, one of the main advocators of parallel universes theory, took issue with their desires to change the theory in order to maintain the one and only world that we live in.[8] He suggested that instead of changing the formalism, why not ask what the formalism really says? In the famous case of the Dirac equation,[9] which was discovered purely by mathematical intuition, physicist Paul Adrien Maurice Dirac found solutions that indicated the existence of another kind of matter, antimatter, that had never been previously observed. By taking Dirac's formalism seriously, experiments later confirmed the existence of the antielectron, the positron, the first presence of antimatter. Thus DeWitt believed that, in spite of the difficulty in observing parallel worlds, one should follow Dirac's example.

Everett, the inventor of the parallel universe theory, in response to Einstein's mouse example, added that from the standpoint of his theory, it was not so much that a mouse could bring on a change in the universe, but the mouse itself was changed by becoming involved with it.

Parallel universes manifest whenever any observation takes place. That means any observation, even if by a mouse, flea, or ameba. When a quantum measurement is completed, a result is recorded in some memory bank—the mind of an observer, as, for example, in the case of a coin in four universes discussed earlier. Although the coin is just a fantasy, it does let me describe in simpler terms how a mouse or a flea differs from a human in the way each observes the universe as seen from a parallel universes perspective.

A Conspiracy of Universes

Remember that this coin has only two things we can observe about it. It has, as physicists say, two complementary observables: *color* and *side.* We can either see that it has a red or green color *or* that it has heads or tails up. But to see a color, you can't tell which side is up; and to see which side is up, you can't determine what color it has.

According to parallel universes theory, the coin is in an indeterminate state until some intelligent being, one who knows the difference between color and side and who is also able to make sense about that difference, comes along and observes the coin. When that person comes along and chooses what to observe, either color or side, the observer and the coin together split and enter into four possible parallel universes:

Universe 1. Heads and the observer sees heads.
Universe 2. Tails and the observer sees tails.
Universe 3. Red and the observer sees red.
Universe 4. Green and the observer sees green.

Now I'm sure you may be wondering about the mouse or the flea and asking what happens when either of them observes the coin. Surely a mouse doesn't really care if the coin has heads up or if it has a red color. If it were a piece of cheese or a cat's whisker—well, that's a cat of a different color, isn't it? The key here is that the observer must be able to tell the difference, and if the mouse can't, then the split doesn't really occur.

Now this answer may seem arbitrary. How does intelligence fit into physics? This is really more of a problem in the one universe—with a consciousness popping a physical measurement into existence—than in the parallel worlds model. But let me explain.

The key idea, the central core of all of the quantum paradoxes, is that possibilities-universes *conspire*. It's a quantum conspiracy. When an observer sees red, he is experiencing the coin in universe 3. However, universe 3 is itself a conspiracy —a superposition of universes 1 and 2. Thus the simultaneous experiences of the observer in universes 1 and 2 *are* what is meant by universe 3. In universe 1 the observer did decide to observe the coin's side. Likewise in universe 2. In each of those universes, the observer intelligently made a choice—to see a side of the coin. In each of those universes, everything is hunky-dory, kosher and logical. These side-universes have no notion of color.

In a similar way, when the observer sees green, and is conscious of universe 4, he too exists as a different conspiracy* of universes 1 and 2. These differences in conspiracy are called *phase differences* by physicists.

Both universes 1 and 2 are necessary. Both exist, unbeknownst to each other. They appear together as the experience of universe 3 (if red is observed) or as universe 4 (if

*The difference in the conspiracies is mathematical. The relationship between universes is found in a mathematical function called the *phase*. In this simple example, this only amounts to whether we add the mathematical wave functions representing the two universes together or subtract one from the other. It is only a difference in phase between universes 1 and 2 that creates universe 3 or 4.

green is observed). It's as simple, in the mathematics of quantum physics, as adding two numbers or subtracting them. Add universes 1 and 2 and you get universe 3. Subtract them and you get universe 4.

In a quite similar way, universe 1 is a conspiracy of universes 3 and 4. So is universe 2—only the phase has changed to protect the innocent.

How can this be, when there are only one coin and one observer? The answer is that we all exist as conspiracies of parallel universes. All our experiences that we say are occurring here and now are also occurring in other universes. Our knowledge of something real and out there gives us the individual experiences we have. The ability to decide what's what and when's when, and where's where—our sense of experience and our sense of will that moves us through space and time with matter—can only result, according to the parallel universes interpretation of quantum physics, when there is a conspiracy, a merging together of the different choices in different universes.

The mouse observing the coin is not split into four mice, because the mouse doesn't have the apparatus capable of deciding what a side or a color is. The mouse sees a coin and that's it. You could, if you wish, think of the mouse and the coin as existing in four universes—but the conspiracy is total for the mouse. The mouse's four universes are one.

Not so for us. We can decide. We do have the measuring apparatus and the ability to make the decision. And when we do, we split. Thus the world we live in, which by now is quite schizophrenic due to past splittings, tends to become even more so as time passes on.

Although Einstein was quite bothered by the mouse's ability to create a parallel universe simply by observing it, he perhaps was a little schizophrenic himself in the fact that he was directly responsible for the discovery of parallel worlds through his insights into a solution of the equations of general relativity he created. It was Einstein who taught us that time and space twist together in a dance that Chubby Checker would have envied. Through this spacetime twist, and its consequent knots, a new path to parallel universes was discovered. That story is told in Part Three.

PART THREE

INSIDE AND OUTSIDE: TIME BENDS AND SPACE WARPS

And space, it has dimensions four,
Instead of only three.
The square of the hypotenuse
Ain't what it used to be.
It grieves me sore, the things you've done
to plane geometry.

You hold that time is badly warped,
That even light is bent;
I think I get the idea there,
If this is what you meant;
The mail the postman brings today,
Tomorrow will be sent.

The shortest line, Einstein replied
is not the one that's straight,
It curves around upon itself,
Much like a figure eight.
And if you go too rapidly
You will arrive too late.

W. H. Williams
At a dinner in the honor of
Albert Einstein on his visit to the
California Institute of Technology,
Pasadena, California (1924)

What did Einstein think about parallel universes? In 1935, he and Nathan Rosen wrote a paper[1] explaining how a parallel universe might be found inside a spherical region of space that was gravitationally highly stressed by the presence of a large massive body at the center of the space. Einstein and Rosen didn't carry their theory too far. Yet their work—now known as the discovery of the Einstein–Rosen bridge—set the stage for the next development in the general theory of relativity. This was the discovery of black holes.

Yet, even though Einstein suspected that relativity theory might predict other universes, he was, as we saw in Part Two, quite unhappy with the predictions of quantum theory. However, in spite of his displeasure, he might have really enjoyed Everett's Ph.D. thesis, had he lived long enough to read it.* Hugh Everett III was a graduate student at Princeton University, where Einstein spent his last days. Everett was the first to consider parallel universes as a serious venture to explain how quantum mechanics and relativity theory could be reconciled. Everett noticed that both theories predicted other universes. Was this a clue to the reconciliation of quantum theory and relativity?

Even today, the problem of consistency between relativity and quantum physics haunts us. The specter of incongruity laughs in our faces. Using our present understanding of physics, we seem to be faced with the most outrageous science fiction. The following questions are only a sample illustrating the realm of consideration now brought forward by the reconciliation of these worlds of modern physics. What may be surprising to the reader is that these questions are now being considered quite seriously by top physicists.

For example, what is inside and what is outside the universe? Does the universe have an energy? Are time and space

*Actually he might have had the chance to read a preliminary version, since Everett completed his thesis in 1957, just two years after Einstein died.

observer-dependent? Do the time and space distortions of relativity imply that one can actually travel in time? If time travel is really possible, what happens if I go back in time and murder my five-year-old grandpa? If parallel worlds exist, what effect do they have on time travel? Can one see beyond time barriers as the old philosophers like Nostradamus used to dream?

One thing at a time. *The* problem in physics today is the merger of the ideas of quantum physics with those of general relativity. In other words, how does one arrive at a consistent view of the universe using the most sophisticated theories invented by the human mind? According to quantum physics, one must be concerned with the actions of an observer whenever that observer measures a property of a physical system. Tacit to this view is the requirement that somehow the observer or observing instrument lies outside or external to the physical system being observed.

However, according to the view of general relativity, the stuff of the universe, matter and energy, are deeply connected to space and time. Thus it is not clear, if we are dealing with the whole universe, what lies inside and what lies outside. Somehow all that counts not only exists in time and space, it also *creates* spacetime itself. Or perhaps it can be thought of the other way round: spacetime *creates* matter and energy. Since space and time are the essential coordinates used to measure anything, a consistent merger must tie quantum physical measurement, observation, together with the view of spacetime as seen in general relativity.

Bringing these theories together is something like solving an intricate Agatha Christie murder novel. Many suspects must be considered. Time and space distortion, matter, the actions of observation, and the mysterious question of the meaning of the quantum wave function. Here lies the deep conceptual problem and a clue. The clue is that both theories predict the existence of the new suspect—parallel worlds. This new suspect on the block appears to be just what is needed to solve all these conceptual problems—and perhaps provide even more surprises.

In Part Three we look at how relativity points to the existence of parallel universes. Our journey, however, takes us through a myriad of new concepts, new ways of thinking about the physical universe. Concepts like imaginary time, imaginary space, time curvature, space curvature, zero-time particles, tachyons, singularities, and others will appear in this part of the book. I don't expect that these new concepts will be easy to grasp. So I've tried to insert a little experiential exercise or example each time a new concept arises. I have attempted to reduce any mathematics to the simplest level I could or put the ideas in endnotes for the reader who has some grasp of mathematics. If the going gets a little rough, you can skip what seems incomprehensible and just move on. You won't need anything in Part Three to grasp Parts Four, Five, and Six. If you do manage to grasp it, so much the better. These ideas are quite fantastic and enjoyable. My point, remember, is that the parallel universes predicted by quantum physics are the exact same as those predicted by relativity theory.

General relativity deals with energy-time-space-matter as conceptually connected. For example, the energy of a system depends on how much of it there is in a given volume of space. But as the space expands to include the whole universe, there no longer appears to be any outside separate from any inside. This is disturbing if one considers the whole arena of spacetime to be closed, as some theories of general relativity indicate.

If there is no outside to the universe, then how could one measure its energy? For that matter, how could there be an observer of the universe, since that observer must stand outside it to measure it? One answer is that there are no observers separate from the universe. Thus the universe cannot be observed unless one can observe oneself at the same time.

To explore these questions tying together quantum physics

and relativity theory, and to see how relativity theory predicts that parallel universes exist, we need to consider the essential elements of any universe as seen in physics. These are space, time, and matter. The special theory of relativity unites two of them into one spacetime and in doing so involves the observer in a previously unsuspected manner.

From relativity we learn that time and space are observer-dependent—they are not universal but relative concepts. This dependency means that two events separated by space and time intervals as seen by one observer will not be the same when observed by other observers moving relative to him or her.

Such relativistic puzzles as the twin paradox (where one twin flies off and back at near light speed to a distant galaxy one thousand light years away and ages only a few years while the other stays at home and ages hundreds or perhaps millions of years so that the home twin is long buried) tell us that the slowing down of moving clocks and the shrinking of moving rulers imply that the clockwork in our supposed single universe is somewhat haywire anyway. Relativity theory says more than that. It says that space without time is not fully meaningful. In some sense they are aspects of one thing only.

Now add Einstein's general theory of relativity. Here the third ingredient, matter, gets enfolded into the spacetime cake. With matter included we find that not only do space and time form a unity called spacetime, but matter is also connected to it. It's like a gigantic sponge cake. The bubbles of spacetime enclosed in the cake are meaningless without the spongy cake surrounding them. You can't even have spacetime without matter. One defines the other, and vice versa.

But matter is more than that. It distorts spacetime—wrinkles it into all kinds of folds and bends, causing time and space to appear quite wacky to a sober scientist. In fact, one cannot decide whether it is matter that is distorting space and time or if it's the distortions of spacetime that are creating matter. It is the old chicken and egg out of the ancient philosopher's basket.

How does matter distort space and time? Through the presence of gravity. Gravity *is* the presence of matter *is* the distortion of spacetime. They are all one and the same thing, according to Einstein. Just a bitty wrinkle in time produces a whopping gravitational force. The earth, for example, distorts a few parts of a nanosecond off the clock when it ticks at your feet as compared to your head. That's enough to hold you on the planet. The more massive the object, the larger the time distortion surrounding it. If a star is really massive, it distorts both space and time equally. It is when such massive distortions occur that relativity theory predicts that holes are torn in spacetime leading to parallel universes.

General relativity theory under the maximum stress of matter's distortion of space and time predicts the existence of a new beast in the world—a black hole, the most distorting region of spacetime anywhere, anytime. And spinning black holes contain bridges connecting our universe to parallel universes. Even a nonspinning black hole has a bridge connecting our world to one other parallel world. Looking inside a black hole we find how such a bridge is built.

If there are black holes in the universe, why don't we see them? In fact there is an argument that this universe is itself a black hole![2] Perhaps all fundamental particles like electrons are black holes. If electrons are black holes (and they do spin), it would seem entirely necessary to include their connections to other universes in order to explain their behavior in this universe. But that is exactly what quantum physics says we must do.

It is the presence of matter and an observer of it that implies the connection between the theories of relativity and quantum physics. Without matter, there is nothing to see. And without an observer, there is no such thing as a universe —for if there were, who would there be to know it was a universe? Both relativity theory when it includes the effects of matter, and quantum theory when it includes the observer, consistently point to the connection between the theories. That connection is parallel universes.

Chapter 11

RELATIVITY AND TIME AS A DIMENSION OF SPACE

I have implied that parallel universes and a new notion of time are conceptually connected through the theory of relativity. In fact their connection might be a clue to their existence and a reason why they are difficult to detect. Our present sense of time may be no more than the track that keeps us in one universe. The discovery of parallel universes will alter our restricted present sense of time and space. To see how, we need to look at what relativity says about space, time, and matter.

Let's begin with time. What is the connection between time and parallel universes? What is time? Having grasped time, what is space? In this chapter we will look at the concepts of time and space as described by relativity. In the next chapter we will see how time turns out to be an imaginary dimension of space, or if you prefer, space is an imaginary dimension of time. In either case, the new view of time expands our awareness capability, making parallel universes more than just a thought.

Primal Time

Physicists regard time as primal, nearly without question. The easiest way to deal with it in physics is to think of it as a dimension of space. Using it this way physicists can describe the movements of objects in a special mathematical language that needs no translation. This universal description of how objects move is called *the equations of motion.* The equations of motion, in this context, need no translation because they are the same for all observers that are moving relative to each other. Hence arises the major idea of the theory of relativity—democracy of description. The laws of motion are the same for all observers if time is thought of as a dimension of space.

However, time is not really a dimension in the normal sense that space has dimensions or that a piece of furniture has dimensions. In fact, in order that time may be used as a dimension in the democratic equations of motion, it must be an imaginary dimension of space.

The First Time I Saw Spacetime

I remember the first time I began to understand Einstein's relativistic description of space and time. Sad to say, this didn't occur when I was an undergraduate student of physics; I was just as confused then as anyone. It came a little later when I received an appointment as a graduate student at UCLA. And even then I didn't really grasp it until I had the opportunity to be a graduate teaching assistant. By teaching it to undergraduates who came to me about their homework problems, I was able to see how simple it was, provided I gave up my commonsense notion about time. It seems that new ideas, like any art form, must be practiced before they really sink in and become part of anyone's thinking.

The key insight for me came visually. I saw, in terms of a geometrical picture, just how space and time could make a four-dimensional *space.*

It is a surprise to me that so little of the new physics is taught today in terms of pictures, particularly when it is being learned. Undoubtedly, much of the controversy dealing with quantum physics and its paradoxes arises from a paucity of pictures. Albeit, there are good reasons why no pictures are drawn. Even Einstein himself did not have one when he constructed his special theory of relativity. He had to turn back to one of his teachers to help him visually understand his own theory. That teacher created the first visual accompaniment to Einstein's spacetime sonata.

The Popular Dimension of Time

This vision, coming straight from high school geometry, originated with Hermann Minkowski. He was one of Einstein's teachers at the Eidgenossische Technische Hochschule (ETH),* in Zurich, Switzerland. Just a year before his death, at the age of forty-four in 1909, Minkowski presented probably the first popular lecture on Einstein's theory in which he used his visuals. He spoke before the Deutsche Naturforscher und Ärzte (the German philosophers and physicians), a body used by scientists to disseminate the ideas of their individual disciplines to a wider audience. Minkowski's popular talk was entitled "Space and Time."

*It was here at the ETH that Einstein failed his entrance examinations. However, Einstein finally received his graduate training in physics at the ETH.

How Time Can Be Viewed as the Fourth Dimension

Minkowski's opening remarks, spoken in 1908, still carry a tone of truth. He said (translated, of course):

> *Gentlemen, the ideas of space and time which I wish to develop before you grew from the soil of experimental physics. Therein lies their strength. Their tendency is radical. From now on, space by itself and time by itself must sink into the shadows, while only a union of the two preserves independence.*

Probably no other scientist more than Minkowski created the Einstein myth. He was largely responsible for the early fame that Einstein received in Germany. Einstein himself described Minkowski's contribution as

> *the provision of equations in which the special laws of relativity take on a new form in which the time coordinate plays exactly the same role as the three space coordinates.*

Now to grasp this, it is important to realize what Einstein had accomplished before Minkowski's contribution.

The Democratization of Space and Time

Einstein had shown that the equations describing mechanical and optical phenomena when seen by an observer moving relative to the phenomena were not the same equations as used by an observer at rest. For example, the equations describing the force between a moving electrically charged particle and a wire carrying an electrical current were quite different if one wrote them from the point of view of moving along with the particle instead of sitting at rest.

For someone watching at rest, the force acting on the moving particle was due entirely to the magnetic field produced by the current in the wire. But for an observer moving along with the particle, there was no force produced by the magnetic field of the wire. The force was entirely due to electrical charges in the wire.

Yet the force had to be the same, regardless of who saw it. Einstein's relativity theory changed the equations and showed how to reconcile the disparity between the descriptions of the resting and the moving observer. The new equations were the same for both observers provided that light had the same speed for all observers regardless of their relative motion. Since the reconciliation dealt with the relative motion of observers and their mathematical descriptions, the name *the theory of relativity* stuck.

The key problem that Einstein had solved by his theory of relativity was integrating the knowledge of electrical and optical phenomena with that of mechanical phenomena when seen by a moving observer compared with a stationary one. Now it was known that an observer moving with a constant velocity could study the movements of mechanical objects surrounding him.

A Relativistic Game of Catch

For example, suppose the observer was in a train moving along at five miles per hour. Suppose he decided to play "catch" with his friend stationed at the opposite end of the railway car by throwing a ball with the speed of seven miles per hour. Since the train compartment is sealed, no gusts of air would spoil their throws back and forth to each other the complete length of the train car. Indeed, their throws back and forth (each at the same speed of seven miles per hour) would appear to everyone aboard the train as just a normal game of catch.

For those stationed on the ground—say, at a railway platform—watching the well-lit railway catch game passing by, the throws would appear different. When viewed as it moved from the front end of the car to the rear, the ball would appear to go slower (two miles per hour). But when the ball was returned it would appear to go faster (twelve miles per hour).

The fact that the ball, traveling at the same speed back and forth inside the car, would travel at different speeds relative to an observer outside the car is not too hard to visualize. It is easy to see when you take the speed of the train car into consideration.

Traveling with the forward-thrown ball, the speed of the train is added to the speed of the ball. Traveling in the opposite direction, the speed of the train is subtracted from the speed of the ball. Thus, the speed of the ball depends on the direction it travels in the moving train when seen from a train platform. But, aboard the train, the ball moves back and forth at the same speed.

If the ball magically turns into a photon—a particle of light—the observations of the photon will not be the same. The speed of the photon will be the same speed, c, according to Einstein, regardless of the direction taken and regardless of who sees it. In other words, both the observer on the platform and the observers on the train will see the photon moving at exactly the same speed, no matter how fast the train is going.

It is not easy to picture this at all. No drawings of baseballs and billiardlike photons will satisfy the reader, not to mention the writer or any other physicist. This is just weird business. It was no wonder that I had much difficulty with it when I was a student of physics. In fact, we would say that this result is counterintuitive. You can't trust your intuition to give you the best answer. The only way to get any handle on it at all is to give up any handle on physical pictures which keep time and space separated. One must begin to redraw things in motion. But the *things* that are redrawn will not look the same. They become objects in four dimensions rather than objects in three moving through time.

What Does a Four-Dimensional Object Look Like?

What does an objection in four dimensions look like? Well, you already know how a three-dimensional object appears. Take a baseball. It looks pretty normal and round covered by a cowhide skin and lace stitching. That is normal when seen in three dimensions. But in four, the baseball looks like a tube stretched out. The extra length of the tube arises in the fourth dimension. The beginning of the tube is back in time when the baseball was first assembled. A bunch of threads are all gathered together finally forming the outline of the ball—the tube's entrance. As the ball moves around it leaves a trail—a long spaghetti winding through time.

Time Is Space Is Time Is . . .

In this very simple conversion, time and space take on similar characteristics. One becomes the other, and vice versa. You may think this is counterintuitive, too. But a little more reflection will show that you already do this for ordinary speeds like car speeds. For example, think of how far it is to the nearest town outside your own. You probably will answer in terms of time, by saying, for example, that it is fifty minutes away. What you are actually doing is dividing the distance between the towns by the speed of your car as it makes the journey at a roughly constant speed down a highway. Thus a town fifty minutes away is probably around fifty miles away, assuming you drive at the unsafe speed of sixty miles per hour, which happens to be the same as one mile a minute.

In fact, this conversion of space into time, and vice versa, using the fixed speed of light as a reference, is now so common to those working in relativity theory that one hardly ever sees a symbol for the speed of light appearing in their equations. If you refer all motion to the speed of light, so that objects move with speeds that are fractions of light speed, then there is no reason to write the symbol c anymore. In this new vision of things, distances and time become highly relativized. For example, one billionth of a second is the same as one foot of length. This is because light will travel one foot of length (actually 11.8029 inches to be nearly exact), in the time of one billionth of a second. Thus a foot and a billionth of a second are the same thing whenever speed is measured in terms of the speed of light. This means that you don't use feet and seconds as standards. You use either feet or seconds, and the other is determined.

Thus the moon is around two seconds away. And in one second you are 186,000 miles away from where you are now. The distance you just traveled was, of course, in time. Just sitting there reading this is quite a long journey space-wise. The sun is around five hundred seconds away and the planet Pluto is about five or six hours away as the light-crow flies.

This vision of space and time together makes our solar system a cozy ten or twelve hours in diameter. Our nearest neighbor star is, however, hardly just around the corner. It is around four years' journey from us. (That is a long time no matter how you measure it.)

However, even though distances and times are the same, for any one observer, they are not the same for two observers moving relative to each other. It was this fact that turned the photon-baseball catch game aboard the train into a paradox. It also turned space and time into spacetime—a whole continuous stream where nothing was moving and nothing was ever standing still.

Chapter
12

REAL TIME, ZERO TIME, IMAGINARY TIME AND REAL SPACE, AND IMAGINARY SPACE

Nothing moving, nothing standing still. This is the paradoxical universe of spacetime. Everything is frozen still. Your whole life history lies like a gigantic centipede stuck in plastic, with one end tiny and babylike in form and the other end old and decrepit. All of your ups and downs are just frozen wiggles in the worm.

As horrid as that may sound, that view made spacetime physics a branch of geometry. So what if it was as horrid as Euclid made it. There was something new and exciting in it. Perhaps all of physics—the stuff of the universe—was geometry too.

And then there was Einstein's democratization of the equations of physics by writing them in terms of spacetime. Whenever it is possible to find a set of equations, as Einstein did, that produce the same physically observable phenomena in spite of observational differences when seen from different viewpoints, physicists say we have an *invariant* relationship. In fact, a discovered invariance is a clue to a deep, previously secret universal connection.

Invariance in Geometry

Now there are many invariant relationships in ordinary geometry. For example, if you draw a semicircle including the diameter (like a half-moon), and then from any point on the circle draw two straight lines to the ends of the diameter, you will find that you always construct a right triangle. (Try it. Take a compass or a large coin and draw a half circle. Draw the diameter—the line going through the center of the circle. Now draw two straight lines from each of several points on the circle going to the ends of the diameter. What do you get?)

All right triangles lying within a semicircle, with their right angles on it, have the same invariant hypotenuse—the diameter of the circle. In ordinary geometry, invariance always leads to novel insights,* such as the fact that the right angles of the triangles all lie on a circle.

*In a standard course in geometry, one usually learns a certain theorem, known as Pythagoras's theorem. It relates the lengths of the sides of a right triangle with each other in a concise and simple formula. It tells us that if one were to erect squares on the sides of the triangle, so that the sides of each square were the same as the sides of the triangle, the sums of the areas of two of those squares would equal the area of the third. The third square is the one erected on the hypotenuse—the side opposite the right angle in the triangle. The formula for this is $a^2 + b^2 = c^2$. Here, the sides a and b are adjacent to the right angle, and c is the side opposite the right angle.

Invariance in Spacetime

But now we are going to use this triangle in a completely different manner by turning our attention to the geometry of *spacetime.* Here, the triangle invariance takes on a new and extremely significant meaning. It tells us, for example, why observers moving relative to each other do not experience the same time passing between the coincident events of their respective lives. But I'm jumping the gun a little. I'll show you why this is the case in a moment.

Minkowski's amazing insight was that the invariance found by Einstein was similar to the simple right triangle invariance found in ordinary geometry. What Minkowski noticed was that if you regard time as a dimension of space, and you construct right triangles with one adjacent side corresponding to time and the other to space, you will find a spacetime invariance similar to the invariance found for the right triangles inscribed in a semicircle. You must make one slight adjustment. You must think of time as an *imaginary* or new dimension of space.

This was the Minkowskian vision of spacetime. Once one gets the idea that space could have another dimension, an imaginary dimension, it opens up all kinds of new visions. It also makes the idea of parallel universes much easier to grasp. They can exist at other locations in the imaginary space. How would these parallel universes appear to us? As ourselves and our universe passing through time.

Now to some readers, playing with imaginative ideas like imaginary space may seem to be so much child's fantasy. I'm sure we have all seen movies or had conversations where someone said that space could have other dimensions or even imaginary dimensions. Such conversations are usually fun and of the "Oh wow, gee whiz" type.

But for physicists contemplating such things, one tempers imagination with a means for carrying out observations.

Thus imaginary space is a useful concept because it leads to a new vision of time—that same time we observe as our passing lives.

Also, as Einstein once remarked, "Imagination is more important than knowledge." In fact, Einstein led the way into a new way of thinking about physics. One first speculates imaginatively, and then suggests what could be observed from such speculations. This, instead of first observing nature and then from experiment trying to deduce the theory.

So for the moment let us "play" with just the idea that space has real and new imaginary dimensions. First of all, what is an imaginary dimension of space?[3]

Imaginary Space

Suppose you walk from one end of the room you are sitting in to the other. You try to walk in a straight line, say, from east to west. Now suppose that you walk in a direction perpendicular to the line you just made—south to north—in your sitting room. I am sure you can imagine that you have just walked off two legs of a right triangle, for you could walk back to your starting point by taking the diagonal path from where you are now standing. In this case, both the legs you walked off that form the right triangle are real—they exist in space.

Now I want you to imagine that your room is a little strange. Distances you walk going east and west are ordinary space distances. But suppose that the distances you walk going north and south correspond to time, not distance. Thus when you walk north you are going forward in time but nowhere in space, and if you walk south, you are going backward in time, but again not moving in space.

I am sure you realize that no matter how fast you made the walk, you not only moved through space, you also moved through time. But just imagine that you are as in control of time as you are of space. Your room would be a kind of time machine. Whenever you walked north, time would advance. Whenever you walked south, time would go backward. Whenever you walked east or west, you would go through space but not move in time. Only when you walked off distances diagonally, say, NE or SW, would you move through space and time together. If you now imagine walking off a right triangle so that the distance you go east and west is one leg of that triangle, and the time you spend walking north and south is the other leg of the triangle, you will have some idea of what a spacetime triangle is.

Now what is the difference between a space triangle and spacetime triangle? The answer is that in a space triangle both legs are in real space, while in a spacetime triangle one of the legs is in imaginary space. I'm sure you can guess that time is the imaginary dimension of space.

Strange as it may seem, the geometry—the mathematical relationship between the legs—of a spacetime triangle is exactly the same as the geometry of a space triangle. The only difference is that the triangle being measured has one leg in imaginary space and the other in real space. The imaginary space was what was meant by the now famous idea that time was the fourth dimension. Thus, you could draw a right triangle in spacetime and it meant something extraordinary.

But the really weird part of all of this was the picture you came up with. By making time a dimension of space, everything was frozen still. There were no moments passing. They were just lines and dots on a spacetime diagram. From here to eternity was just a line.

A Change in Time

It is here in all of this imaginary and real spatialized passivity that relativity shows how time is changed. Just as the distance traveled by a city dweller in going from street to street varies according to how he crosses those intersections, the same can be said for the time spent by a spacetime traveler. Two spacetime travelers could travel from one place to another and both arrive simultaneously, as seen by a third observer. Yet each could take different times to do it, depending on their speeds.

As fantastic as this sounds, many experiments with atomic and subatomic objects have confirmed these bizarre results. Thus time is relative: it depends on the relative speeds of the time observers. This is just another way of saying, it depends on how you construct your spacetime triangle.

So long as the imaginary, or time, side of a spacetime triangle is longer than the real side, the hypotenuse will also be an imaginary or timelike side.* But what happens if the imaginary side is the same length as, or shorter than, the spacelike side? What would that mean? To grasp this consider the following:

Going Down Fifth Avenue in Spacetime

Let's assume that New York City is a city in spacetime, not just in space. Going down Fifth Avenue (where the traffic usually flows faster) corresponds to moving in imaginary space (time) but not moving in real space. Going crosstown

*Here a little explaining will be useful to those of you who have some mathematical grasp. Remember that everything that moves is gauged in terms of the speed of light. Thus if one asks how long a time leg of a spacetime triangle is, it, of course, depends on the amount of time spent. That time is measurable in seconds, minutes, years, milliseconds, microseconds, nanoseconds, etc. You choose a scale of, for example, seconds, and the time leg is longer, the longer the amount of time spent in seconds.

But distance is no longer thought of in terms of feet, yards, meters, or miles. It is referred back to the speed of light. Thus if the time leg is gauged in seconds, the space leg is gauged in the distance that light could go in a second. This is called a light-second. If the time is gauged in years, the distance is gauged in light-years.

Thus a spacetime triangle with a time leg longer than a space leg means that the object traveled more seconds than light-seconds. In other words, it went farther in time than it did in space. This means that it moved slower than light would have.

along Fifty-second Street corresponds to moving in real space, but not really moving in imaginary space (time). And jaywalking corresponds to moving in both spaces simultaneously.

Now it is perhaps too confusing to think about moving in space and moving in time separately. In real life, as we experience it, we never can move in space without taking time to do it. And most of us can barely imagine moving in any space, real or imaginary, anyway. But moving in time without moving in space? Well, actually, it is easier to experience moving only in time than it is to imagine it. You are doing it right now. In fact, moving in time is what we call sitting still. As we sit, we say time marches on. Actually time doesn't do that. We march on in time. In fact, from the point of view we are taking here, we march on through an imaginary dimension of space called time, moving at the speed of light.

An Ordinary Particle

Now that is a big statement, so let's consider some examples to see what it means. First, suppose that a New York City spacetime dweller travels crosstown a distance of three blocks and uptown a time of five microseconds. A microsecond is one millionth of a second. Suppose that a crosstown block is about three hundred meters long. (It really isn't that long in real Manhattan. It just seems to be when rush hour has started and it is raining.) Now since this is spacetime Manhattan, remember just as in your imaginary room, movement up and downtown corresponds to movement in time and not in space, while crosstown traffic moves only in space but not in time. Since light moves at the speed of one block per microsecond, our traveler has just walked crosstown a distance of three blocks or three light-microseconds —the distance that light goes in three microseconds.

Our dweller, since he has gone five microseconds uptown as well, has also just walked off both legs of a spacetime triangle. We then predict for him a speed of three-fifths the speed of light—the ratio of the lengths of the sides of the spacetime triangle.

Suppose he was born at the starting point and died at the finishing point—a very short life, I agree. The surprising thing is that the hypotenuse of the triangle turns out to be a time—an imaginary space. In fact, it turns out to be the time he spent as he observed it. His lifetime as seen by himself is the length of the hypotenuse of the triangle. Consequently, he would actually live for only four microseconds, even though his lifetime as seen by city dwellers that never went crosstown in spacetime Manhattan was seen to be five microseconds.

So much for a person who lives longer in imaginary space than in real space, or, in other words, travels more in time than in space. Such a person will turn out to always travel slower than light speed, spending his life in longer imaginary space (real time) and shorter real space. Physicists call particles that behave this way, *bradyons*. The root word *brady* means slow. Thus a bradyon is a slow-moving particle.

A Zero-Time Ghost

But now consider a person who travels crosstown three light-microseconds (that is, three blocks) and uptown three microseconds. Suppose that he was born at the start and died at the finish. Since three microseconds are a distance in imaginary space, what can we say about this speedy fellow? Well, he was observed to live for three microseconds and to travel three light-microseconds. But his own self-observation was that he didn't live at all! He existed for zero time* (the hypotenuse of the spacetime triangle).

*Here you must use the Pythagoras relation, $a^2 + b^2 = c^2$. The imaginary leg of the triangle has length, $a = i3$, and the real leg has a length, $b = 3$. Using the triangle relation, $(i3 \times i3) + (3 \times 3) = -9 + 9 = 0$. Thus, the hypotenuse has length $c = 0$.

This is a bit shocking. Did he exist or not? Could there ever be a particle that behaved this way? By the way, according to relativity, he didn't go anywhere either, since according to his own reckoning, he spent no time. Now it may surprise you that zero-time particles do really exist. In fact we see them every day. Not just in physics laboratories, although there are certainly a wide variety of them there too; but in our everyday lives. They are, in fact, so common that we nearly completely ignore them, unless they happen to not appear at all, and then we get downright worried. These are particles of light, the same light we see with our eyes and that physicists call photons.

When I first realized that photons spend no time in our world (or in any other world as far as I can tell), I was shocked, amazed, and delightfully mystified. Here we have biblical mysticism supported by modern physics on a grand scale. Not only was the universe created from nothing, as we will see in Part Four, and in the beginning there was light out of which matter was created, but light, like a disinterested sculptor, spends no time in the universe that it created.

We see light. We calculate how long it takes for a photon to travel from one place to another. It travels one light-second every second, one light-hour every hour, and one light-year every year. It always goes at the speed of light and therein lies its charm. It can't go anywhere and it can't even exist here for an instance, according to its own way of reckoning.

For light, birth and death are one. This property of zero-time particles makes photons exist on a borderline between solid, tangible and the ethereal, potential. Photons are all of these simultaneously.

Thus a particle that spends equal amounts of time in space and in imaginary space must always travel at light speed. These are zero-time particles. But zero time means zero imaginary space, and since zero is zero, it means zero real space, too.

An Imaginary Time Superman

Carrying on, suppose that a person goes uptown three microseconds and by sheer courage crosses town a distance of five blocks or five light-microseconds. We then draw a spacetime triangle with its real leg longer than its imaginary leg. What kind of weird person would this triangle represent? Now this person will not have a zero-time hypotenuse. In fact, his hypotenuse won't be a time—an imaginary space interval either. His hypotenuse will be a real space interval—a real length of four blocks or four light-microseconds.*

Just as imaginary space meant movement in time without movement in real space—in other words, simply standing still and moving in real time—what could a real space hypotenuse mean? The answer is movement in real space without any movement in real time. We call such a movement, *imaginary time.* Thus this person, although he is observed by a stationary observer to move in both space and time, i.e., in real space and imaginary space, travels by his own "wristwatch" only in imaginary time. Imaginary time is the same as real space. It is seen by himself as movement in space alone. If we determine the speed of this person, it comes out to be 5/3, or one and two-thirds the speed of light. He goes faster than light! Look up in the air, it's—you know who.

Now Einstein showed that faster-than-light particles could not be created because it wasn't possible to accelerate a particle from rest to the speed of light. That's true. It would take an infinite amount of energy to do so. But a particle already existing with a speed greater than light speed is not disallowed by relativity. It's just another weird situation. Physicists call these particles *tachyons.* The word *tachy* means

*Again use the triangle relation $a^2 + b^2 = c^2$. Here $a = i3$, $b = 5$, and so we get $(i3 \times i3) + (5 \times 5) = -9 + 25 = 16 = 4 \times 4$. Thus the hypotenuse $c = 4$.

fast, as in tachycardia, meaning fast heartbeat. In this case, a tachyon means a particle faster than a photon, a speed greater than light speed.

Thus tachyons are particles that move in imaginary time as seen by themselves, just as photons are particles that move in zero time, and bradyons are particles that move in real time. Imaginary time particles do not experience time the same way that we do. Remember, we experience real time, which in relativity is the same as imaginary space.

Now by analogy, since imaginary space is real time, imaginary time must be experienced as real space. Thus a tachyon would be able to move in imaginary time just as bradyons move in real space.

Moving a bradyon in real space (imaginary time) is not a big deal. Just get up and go for a cup of coffee. Are you back in your cozy chair yet? There, you've just moved a bradyon (your body) in imaginary space (real time) and in real space (imaginary time). Notice how easy it was to go backward and forward (from the chair to the kitchen and back) in imaginary time. Not so easy in real time, I'm afraid.

Tachyons do not experience real time. They are thus able to move backward and forward in their imaginary time dimension as easily as you go back and forth from your reading chair. But we would experience tachyons in real time.

Chapter 13

EINSTEIN'S PARALLEL UNIVERSE

The insight that Minkowski gave to Einstein's theory of relativity brought forward a new and perhaps mysterious insight into the workings of nature. It appeared that it may be possible to derive the properties of physical substances in terms of geometry—provided you were willing to take imaginary numbers into account. This won't be the only time this happens in physics. Imaginary numbers are also needed to make a quantum physical theory of matter.

This put all space and time measurements performed on real objects, moving or at rest, on the same footing as geometrical measurements using nothing more than some kind of yardstick. The yardstick was, of course, figurative. It had to measure the sides of triangles in spacetime.

However, Minkowski's insights into spacetime geometry were based on the ideas of Pythagoras and Euclid, who were mainly concerned with space that was as flat as the earth apparently was in their day.

If Pythagoras Is Right, Spacetime Is Flat

What made the geometry of Euclid and Pythagoras useful was that it is always true if the surface containing the figures they drew was as flat as a pancake rather than curved like a baseball or a saddle. Thus Euclid's relations and the triangle formula of Pythagoras were valid for the flat earth but not for a curved planet. Another way of saying this is to speak of figures drawn on a flat rather than a curved space. Figures in flat space obeyed the old geometry. Figures in curved space did not.

Surveyors use right triangles and the old formula of Pythagoras every day of their surveying lives. Suppose they wanted to measure the distance across a pond. By measuring the distance along one shore and the distance along another shore that was perpendicular to the first shore, they could construct a right triangle that had the hypotenuse running across the pond where they could not measure. The two legs of the triangle would give them the distance across the pond with the aid of Pythagoras's formula. The pond is small enough that the curvature of the earth does not distort their measurements. However, if the surveyors were to measure across an ocean, this formula would not be correct. The ocean lies on a curved surface whose curvature is on the same scale as the distance they measure.

Thus to test a space, you draw a right triangle in it and see if the Pythagoras theorem holds water. If it does the space is flat, if it doesn't the space is curved. If the space is curved, the formula will not be true.

What happens to a spacetime triangle when spacetime is curved? The answer is that old formulas do not hold because matter is present. And if matter is present, time and space are distorted. And if they are distorted enough, holes in spacetime appear. And through those holes we find parallel universes. Here we will explore how Einstein came to this conclusion.

Visualizing a Curved Space

To begin we will look at what is meant by curved space. A curved space may seem strange at first, but it too is based on our ordinary common experiences with the world we live on. I say *on* because the planet we all inhabit is a prime example of a curved space. It is as near to a sphere as practically possible. And on it the flat space triangle relation no longer holds.

It is useful to see the difference between a triangle drawn on a sphere and one drawn on a plane. It will help us to grasp what takes place in a curved spacetime.

First of all, imagine taking a journey from the north pole of the planet, making your way down to the equator, walking along the equator for a few miles, and then turning northward, returning to the north pole once again. You have just walked off a spherical right triangle. In fact, if you think about it for a moment, you can see that the triangle you walked doesn't quite match up with a right triangle in flat space. An obvious difference is that the spherical right triangle actually contains two right angles while the flat one has only one. Furthermore, the two legs extending from the equator are parallel when they start out, but cross when they reach the north pole. (You might remember that in flat space geometry, two parallel lines never cross.)

And, of course, since the triangle actually lies on a rounded surface, not a flat one, any formulas relating to it will change. But mathematicians aren't particularly bothered by this. The thing that they are most concerned with is distance. They want to know how far it is from one point on the globe to another. Since there are many ways to travel on a globe, they also want to know what is the shortest distance possible between any two global points. (So do you, if you ever take a sailing or air trip.) They want to determine what relationship holds between the "legs" of a spherical triangle and its hypotenuse. Sailors and pilots determine the distances they travel by using instruments to correct for map errors. They often use navigational aids, such as the loran system, that enable them to get a "fix" from satellites overhead.

Now there are two perpendicular lengths that one can draw that lie totally on the surface of the sphere. These lengths are also familiar to you: They are the lines of longitude and latitude. The lines of latitude are perpendicular to the lines of longitude. What's strange about them is that each line, when extended, forms a perfect circle. The longitudinal lines all run through the poles. Each longitudinal line circumscribes a great circle. Each circle has exactly the same circumference, with its center at the center of the sphere. The latitudinal lines all circumscribe the globe running east-west (or west-east, if you prefer). They do not have the same circumferences, nor are their centers at the center of the globe. The centers of the latitudinal circles all lie on the axis of the sphere running from the north to the south pole.

(Contrary to a flat plane where the perpendicular lines extend to infinity, lines in a curved space often come back on themselves. We will see a little later how curved lines in spacetime manifest as gravity.)

The fact that the latitudinal circles are not the same circumference changes the geometry and any of its relationships. Instead of a simple relation between a length running diagonally and the perpendicular legs of a triangle,* there is change of scale associated with distances running latitudinally. This scale alteration is due to the change in sizes of the latitudinal circles.

For example, suppose you wanted to travel west to east and you wanted to determine how far you went. Consider a

*Such as the flat triangle formula, $a^2 + b^2 = c^2$.

journey that circles the globe at a constant latitude of 45 degrees north. (This journey would take you across the northern continents. Starting in Portland, Oregon; through St. Paul, Minnesota, and Halifax, Nova Scotia; across the North Atlantic, through Bordeaux, France; Venice, Italy; Bucharest, Romania; across the Aral Sea in the Soviet Union; across the Gobi Desert in Mongolia; the northern tip of Japan below the Aleutian Islands; and back to Portland again.) The total distance traveled would be proportional to the complete sweep of 360 degrees of the circle. In fact, it would be equal to the circumference of the circle you journeyed.

If you compare this trip with one going around the equator, the equatorial trip would be longer. The difference in the lengths is the scale factor mentioned above. It depends on the latitude itself. A 360-degree journey at 45 degrees latitude is about 70 percent of a 360-degree equator journey.

Suppose now that you wanted to get from Portland, Oregon, to Bordeaux, France, along the quickest and shortest possible route. Is the previous journey, at a constant latitude of 45 degrees, taking you through St. Paul and Halifax, also the shortest? The answer is no. It turns out that the shortest journey must deviate from a path of constant latitude. The shortest path is a segment of a great circle, one that has its center at the center of the globe.

It actually goes northward, crossing the middle of the Hudson Bay at 60 degrees latitude; passing over the town of Povungnituk on the bay; the Ungava Bay; a little farther northward over Tingmiarmuit, Greenland; then heading slightly southward over Cork, Ireland; and finally reaching Bordeaux. The total distance is about 75 degrees of a great global circle (around 5212 miles). Going along a constant latitude line uses up around 90 degrees of the arc of a great circle (6255 miles). It is easy to see this if you have a globe. Take a piece of string and stretch it around the globe placing one end of it at Portland. Now try to find the shortest string length going from Portland and reaching Bordeaux, keeping the string on the globe. That length is a segment of a great circle, and you will see that it follows the course I mentioned fairly accurately.

Finding the shortest route between two points in a space and comparing it with the shortest route in a flat space tells you how much curvature you have. The shortest route possible in any given space, such as the surface of a globe, is called a *geodesic*.

In flat space, geodesics are straight lines. You have already heard many times that the shortest distance between two points is a straight line. Indeed this is so, provided you are talking about a plane, a flat surface, on which that line exists. On flat surfaces, straight lines are geodesics—the shortest distances between points on the surface. On globes they are arcs of great circles, as I described above. In other spaces they may not be so easy to visualize. Introducing curvature to spacetime will make the geodesics difficult to visualize, but easy to experience. Those experiences are known as "free falls," and our astronauts, circling the globe miles above our heads, experience spacetime geodesics as everyday parts of their lives.

When an astronaut circles the globe several hundred miles above the surface, he is actually experiencing no gravity at all. He is in a state that is called "free fall." His path through spacetime appears as a spiral, with the axis of the spiral corresponding to the time dimension and the radius of the spiral corresponding to the space distance between him and the earth. The shortest distance between two points around the earth is a track following that spiral.

As he approaches the earth, the spiral tightens. As he moves farther from the earth, the spiral lengthens. Thus the spacetime is curved and the curvature varies.

Einstein's Curves

I seriously doubt that Einstein would have thought of using geometry to explain gravity without the insight into spacetime brought forward by his teacher, Hermann Minkowski. Minkowski made Einstein realize that time could be used as imaginary space (which made spacetime no more difficult to deal with than flat ordinary space) so long as spacetime was flat and Pythagoras's theorem held. This gave Einstein an insight.

Suppose spacetime was not flat. What would curvature of spacetime introduce? Well, it turned out to introduce a lot. It gave gravity a new name and a new geometrical significance. It also made physics nearly incomprehensible, even to those who had enough background to understand what Einstein was up to.

Einstein was aware of the bewilderment introduced by attempting to "picture" physics in terms of geometry. The concept that all of our lives are merely lines on a spacetime map, with the various events of our lives no more than points on those lines, made most people feel quite insignificant. This probably caused more rejection of Einstein's ideas than we might imagine.

He once wrote[4]:

> *The non-mathematician is seized by a mysterious shuddering when he hears of "four dimensional" things, by a feeling not unlike that awakened by thoughts of the occult. And yet there is no more commonplace statement than that the world in which we live is a four-dimensional continuum.*

So I do appreciate the bewilderment most of my nonscientist readers experience when they realize, as Einstein did, that physics had passed from the merely observable and physically tangible into the world of the mathematician. Einstein, himself, had trouble with the mathematics he was creating. He usually worked intuitively and sometimes saw beyond his own horizons.

Einstein Going Around the Bend

What gave Einstein his mathematical insight? Probably his realization that gravity could not be explained in flat spacetime—a spacetime where Pythagoras's theorem held.

Now gravity doesn't seem to be too difficult to explain. What could have been on the mind of Einstein? I mean, isn't it just a big magnet holding us all on the earth? Or has it something to do with the earth's rotation? Maybe if the earth stopped rotating we would all fly off into space or something. In fact, this idea was used in a film, *The Man Who Could Work Miracles,* with some success.

We all experience gravity every time we step on a scale. Our weight reassures us that we are present on this whirling planet. I'm sure that we have worried about our weight a little from time to time. But weight really isn't the problem. We would weigh nothing whirling in space as the astronauts do. Or if we stepped on the moon carrying our trusty scales with us, we would weigh one-sixth what we do here.

According to Newton, gravity is an invisible force that exists between all bodies. Every body in the universe attracts every other body with the force of gravity. Gravity holds us to the earth and it holds the planets in orbits about the sun. It holds the moon and the earth together as partners, each whirling about their common center of gravity. But this wasn't what Einstein had in mind. He wasn't thinking in a straight line. His mind was on curves.

He was undoubtedly worried about a little problem in attempting to fit gravity into a geometrical picture. That problem was what happens when an observer is accelerating with respect to another observer. He knew that, at least locally in one's immediate environment, the effects of gravity could be canceled out by an acceleration in the right direction. Indeed, astronauts circling the globe feel no gravity at all because they are accelerating around the world, circling at around eighteen thousand miles per hour.

He also knew that the effects of gravity could be experienced in the same way as undergoing a constant acceleration. A familiar example is to picture yourself standing in a completely soundproof elevator with no doors or windows. The elevator is attached to a rocket and the rocket is firing, causing it to accelerate upward. The elevator is in free space, away from the earth. You will experience the floor pushing up on your feet with a force. If you drop an object, it will stand still, but the elevator's floor will accelerate upward and hit it. With a little thought, you can see that the effects of the elevator's upward acceleration are the same as if it existed in a gravitational field back home on the earth.

This equivalence can be visualized by returning to the curved space of our planet's surface once again. To make this clear, we need to grasp what acceleration means as a spacetime picture. When an object moves from one place to another at a constant velocity, it can be pictured to be moving through space and time following a straight line. This line can be drawn on a flat sheet of paper. Now think of time, once again, as a dimension of space. The piece of paper is flat and can be thought of as a two-dimensional space—in other words, a surface. If that sheet of paper is in front of you, then any line going up the page from the bottom corresponds to the time dimension and any line going across the page corresponds to a single spatial dimension. The lines cross making a ninety-degree angle. This is a flat spacetime map.

Going back to an object moving at constant velocity again, suppose that the line representing the movement of the object also goes from the bottom to the top of the page. This corresponds to the object moving through time but going nowhere in space. This is an object at rest. If the line made by the object is drawn diagonally across the page, the object is moving linearly through both space and time. This is an object traveling with constant velocity.

But if we drew a curved line on the paper, the line corresponds to an object undergoing acceleration. Since acceleration is equivalent to gravity, and curvature of a line is equivalent to acceleration, then curvature is equivalent to gravity. Thus reasons Einstein.

Now imagine journeying from the equator toward the north pole. A few miles away along the equator is a friend. Each of you is going straight northward—as straight a line as each of you can make on the curved two-dimensional space of our planet. Even though you are each marching along following geodesics you will begin to notice that you are gradually coming closer and closer together as you both approach the north pole. You are traveling, not in the "twilight zone," but on a curved space. And, in a sense, accelerating toward each other because of the curvature. You might even say that there was a "force of gravity" pulling you to each other, if you didn't know that the surface of the planet was curved.

In just this sense, we can think of gravity as curvature of space. This curvature is, however, in both space and time, not just in space alone. Because we move relatively slowly on our planet, the gravity we commonly experience on planet earth is more a curvature in the time dimension than in the space dimension. In other words, gravity can be related to a time warp—a distortion in the movement of time as one moves from a higher to a lower room in a building.

Measuring Time Warps

Can we measure the difference in time between two locations? The answer is yes. This is called the *gravitational red* *shift,* and its effects on clocks were first measured* by physicists R. V. Pound and G. A. Rebka at Harvard University in 1959. The red shift is the change in color observed from a radiating body that is moving away from us. Color is our way of noticing the wavelength or frequency of a light wave. A shift in color making it redder means the same thing as a shift in frequency making it lower. Glowing bodies appear redder when they move away from us and bluer† when they approach us. A similar thing is heard when a train approaches us and then recedes from us blowing its whistle. As it approaches us, the frequency appears higher and when it recedes from us, it appears lower.

The same kind of thing takes place when a glowing or radiating object is placed in a gravitational field. A radiating object glows redder when it is put in a stronger gravitational field and bluer when it is placed in a weaker field. Similarly, anything that vibrates shows a shift in frequency when placed in a gravitational field. Clocks tick slower in a strong gravitational field—are red shifted—and faster in a weak gravitational field—are blue shifted. It was the first measurement of a time warp. Pound and Rebka set up two timekeepers, one in the basement of a building on the Harvard campus and the other seventy-four feet above in the building's penthouse.

According to our geometrical picture, the two clocks would not keep the same time because of the spacetime curvature produced by the gravitational field. Next the two physicists figured out how to send a timing message from the basement to the penthouse. This message would carry a very small time interval in it. This time interval matched the clock rate in the basement but, because of the red shift, would not match the clock in the penthouse.

*Pound and Rebka's experiment is discussed in *Gravitation*, by Misner, Thorne, and Wheeler. *See* the Bibliography.

†The change in color is apparent only for objects moving at very high speeds.

How large was the time warp measured by Pound and Rebka? The effect was quite tiny; indeed it was amazing that it was even measured at all. The result was a slight increase in the length of one second of time (a mini-time distortion). In terms of one second, this was a very small change. When compared with the clock just seventy-four feet above it, the lower clock's second lasted just about one divided by thirty-four powers of ten longer.

Although this time warp is extremely tiny, it is the cause of our gravity. The powerful pull holding us to the planet is this tiny time warp. The greater that time warp becomes, the stronger the gravitational pull on us. This experiment confirmed the slowing down of time as predicted by Einstein's geometrical spacetime curvature theory.

How did that red shift indicate that time had slowed down in the basement? As I pointed out earlier, if there is a gravity field there must be a difference in time curvature. Since our gravity points downward, the time curvature at our feet must be slightly greater than the time curvature at our head. The higher up we go, the less the time curvature. The farther down we go, the greater the time curvature and the slower the ticking of clocks and all physical processes. Go up time speeds up; go down and time slows down. We all walk around this planet oblivious of the fact that it is a time machine. That means that clocks in the basement will run slower than clocks in the penthouse.

A Bend in Time Means a Parallel Universe

Let me briefly remind you of the direction we are traveling in. We are attempting to connect all of this Einsteinian "stuff" with parallel universes. The bridge between gravity and parallel worlds comes through a solution to Einstein's mathematical equations for a curved spacetime. Curvature in the equations introduces surprises. These surprises appear as what mathematicians call *singularities.* These are regions of spacetime where gigantic distortions, possibly even rips in the fabric of spacetime, appear. At a singularity, all physical quantities take on infinite values. These singularities exist at the centers of objects called *black holes*. In these black holes, space and time become highly stretched. Time intervals stretch so much that light slows down to zero speed as it approaches one. In other words, light stops moving at the center of a black hole, and the laws of physics go crazy. In the vicinity of these distortions there are gateways to other universes. The first to discover one of these gateways was Einstein and Nathan Rosen. It is now even called the *Einstein–Rosen bridge*, and it connects two different parallel universes. We'll see how in the next chapter.

Thus if there is gravity, there is curvature of spacetime. And that means there are, as far as we can tell today, black holes, gateways to other parallel universes.

Chapter 14

BLACK HOLES: GATEWAYS TO PARALLEL UNIVERSES

In the last two chapters we saw how a geometrical view of spacetime provided an insight into the structure of the universe; how time can be viewed as an imaginary dimension of space and how a curve in spacetime can be seen to be the cause of gravity. It is remarkable that we live in a universe which, although appearing quite ordinary for most of our everyday lives, nevertheless has at its core a structure that bends time and space.

We have accepted gravity as a natural force in the universe, holding us to our planet's surface and holding up the stars in their orbits about their galactic centers. Yet all of that is nothing more than straight-line motion in curved spacetime. We all move along such "straight" lines (called geodesics) completely oblivious that the spacetime we move through is highly curved. The only evidence of the curvature we see or sense normally is the gravity that holds us to the earth.

If the objects under consideration are moving slowly compared with light speed, they will experience gravity as mainly due to curvature in time. It, more than space curvature, is the cause of ordinary gravity. In other words, what we call gravity as we experience it here on earth, is the actual experience of living in a time warp.

Space Warps

Is there a space warp, too? The answer is yes, because time and space are united. But we are hardly aware of space warps. Why do we not experience the space warp as much as we do the time warp? The answer is that we move too slowly through space. Or in terms of our geometrical picture, we move through time much, much faster than we move through space.

How fast do we go through time? The answer is, at the speed of light.

This last statement can be taken in several ways. If you wish, you can think of yourself as being totally at rest. (Indeed, you are probably at rest reading this right now.) You feel this "rest" because nearly everything around you is sweeping through time at the same rate as you are. Relativity explains that whenever you move at a constant speed through spacetime you experience the world normally. One second takes one second. One meter stick is the same length as another. It's only when there is relative movement that we see things any differently.

Thinking of time as an imaginary distance means that each second of time we are actually moving through imaginary space. Using the speed of light as a scale, we measure this distance in terms of how far light travels. We use light as a scale because it is the only constant speed in the universe. Thus a movement of one second in time is actually a movement of 300,000 kilometers (186,000 miles) in imaginary space. Thus vast distances exist between each second—the distance that light travels in that second.

Using this as a guide as we spacetime travel through our ordinary lives enables us to grasp how difficult it is to turn around in time. It is like attempting to turn around on a highway when you are rushing along at a great speed. Inertia alone carries you along the highway.

This is the reason that we find it difficult to detect space warps. We are not moving far enough or fast enough through space to pick up any deviations from a straight line. It's similar to taking a walk in your own neighborhood. Except for an occasional hill, the ground around you appears quite flat. If you happen to be near a great body of water, you may peer out at the horizon. It even looks perfectly flat. Yet it is curved—following the curvature of the earth. Often on a clear day we can see the earth's curvature. However, it is only from a new vantage point (as, for example, gained by our astronauts) that space curvature is really seen in all of its glory. The earth is clearly seen to be round.

Even though the space curvature of the earth is apparent, the space warp produced by the earth is practically invisible. It does appear as a slight difference in the gravitational pull acting at our feet as compared to the pull acting at our head as we stand on the surface of our planet. This difference is very slight because the earth contains little mass when compared on a cosmic scale—the scale where black holes become important.

If we could imagine the earth undergoing a shrinkage—what physicists call a collapse—things would begin to look a little different. The space warp would begin to make its presence apparent to us.

How Space Curvature Creates a Black Hole

Our planet is massive. It contains about 6×10^{24} kilograms of matter. That's a vast figure. Six million billion billion kilograms. Yet, as massive as it is, the earth is a tiny hunk of matter, cosmically speaking. The sun is much more vast. It contains around 333,000 times the mass of one earth. This is called one solar mass. Our galaxy is even vaster. It contains around two hundred billion solar masses of matter. In comparison our planet is but a speck.

On this scale, our planet is not very large either. It is approximately 12,000 kilometers in diameter. Compared to the sun we could stretch nearly 110 earths from one solar pole to the other.

Even though it has a "tiny" mass, we could imagine our planet, under very special circumstances, being squeezed into a much, much smaller region of space. Under these circumstances a space warp surrounding the earth would appear. What would we observe if such a squeeze occurred?

To see this, let us imagine that the earth started to shrink, but contained its same mass. Although this may be difficult to imagine, just remember that all atoms are mostly space and are separated by space. By squeezing matter together we are taking away some of that space but keeping in all of the matter. Suppose, at first, the earth was shrunk to one half of its present radius. The first thing we would notice is that the force of gravity would increase by a factor of four. This means that each of us would weigh four times as much as we do now.

However, the force of gravity at our feet would still be roughly the same as the force at our heads. Thus the time curvature would be four times as strong, but the space curvature would still not be too apparent to us.

Continuing to squeeze the planet down we would find that every time we halved the radius, we would increase the gravitational pull by a factor of four. But more important as far as space curvature is concerned is the fact that we would also begin to feel as if we were being pulled apart by the growing difference in the force of gravity at our feet and at our heads (assuming that we are standing). This difference in force is called the *gravitational tidal effect.* If we could continue this squeeze until the earth had a radius of only around one half of a centimeter, something very new would begin to manifest. The earth would become a black hole. If the earth were to be squeezed to a point, the difference between the force of gravity at our feet and at our heads would become infinite.

Suppose that someone on the moon was watching all this happen. His observations would also seem very strange. Just before the earth shrank to a critical distance of about .4438 centimeters, the spacetime surrounding the planet would become so distorted that any object that fell toward the earth would appear to never ever reach the planet's surface. In fact the movement of the planet's surface as it was collapsing from just above .4438 centimeter to just below that radius would seem to take forever. Objects falling toward the earth would also appear to be journeying slower and slower as they approached the planet's surface.

What Is a Black Hole?

The distortions of space and time surrounding a planetary object are predicted by Einstein's general theory of relativity. Although it is not very likely that a big earth squeeze would occur at any time in the future (because the earth's mass is too small), the possibility is quite real for large stellar objects, such as stars with masses just over three solar masses (stars three or more times as massive as our own sun). These stars are just too massive. They cannot stand against the tides of gravity. Even though they are massive enough to become thermonuclear furnaces and radiate away some of their mass, turning it into radiant energy, they will eventually suffer the consequences of being overweight—they will turn into black holes.

For every black hole that forms, there is a certain critical distance, called the Schwarzschild radius or gravitational radius, which marks the division between the outside and the inside of a spherical surface defining the outline of the black hole. Really a black hole is a misnomer. A black sphere is more like it.

To gain some idea of what a black hole looks like, suppose that we were standing on the surface of a distant planet watching an overweight "sun" as our planet orbited about it. Suppose that the sun was just about to go black—turn into a black hole. To make it interesting, suppose that a fellow astronaut stood on the sun's surface and sent us messages about the collapsing star just as it began to shrink to the critical radius where it would suddenly go black. If he sent messages every second, so that each message was uniformly spaced by exactly one second, these signals would propagate to us at the speed of light. However, as the star crossed over the critical radius, these messages would begin to distort. Each message would take a longer and longer time to reach us. We would have to wait forever to see the star turn black.

At first the signals would arrive each and every second. But as the sun continued to collapse, the messages would start to arrive later and later—every two seconds, then every four seconds, and then every minute, then every hour, then every day, then every year, then every century, then every millennium, and finally we would see no messages at all. In fact, we wouldn't ever see the star go black. It would take forever for it to turn black.

This is what a black hole forming would look like for a distant observer. A star taking forever to go out. And the light it sent would become redder and redder with each passing moment. It would look like a dying ember in a fireplace.

But for the astronaut on the sun's surface, nothing would seem out of kilter. He would still send messages every second, as far as his clock would say. He wouldn't notice that the spacetime surrounding him, his environment, was distorting time. Indeed, he would find that the collapse was taking a finite amount of time to occur.

If he were careful, however, even though he sent his messages each and every second, he would notice some weirdness. His messages would be uniformly spaced in time, but they would take longer and longer to leave the star. Light speed would appear like molasses flowing. Just when the star collapsed below the critical radius, all messages would actually turn around and be sucked in by the star. The star's gravity would be so strong that even light would be pulled back in by it.

Clocks on the star's surface as it collapsed would be running more and more slowly in terms of clocks on the orbiting planet. The star would appear to be collapsing forever, and take forever just to reach the critical radius.

It is this sudden appearance of forever in the equations that makes black holes such unusual objects. The key is the appearance of the critical radius, the boundary of weirdness surrounding the collapsing star. When the star collapses to that critical radius it becomes a black hole.

Chapter 15

HOW BLACK HOLES PREDICTED PARALLEL UNIVERSES

Thus it was that Einstein's theory predicted the existence of such exotic objects as black holes. However, by 1935, Einstein and his associate, Nathan Rosen,[5] at Princeton University, realized that the black hole had another surprise. The hole wasn't a hole after all—it was a tube connecting onto another possible universe. As Einstein and Rosen first stated it, there was a "bridge" through a black hole to anywhere and anytime,* and another younger Princeton physicist, Martin Kruskal, was the first to map it.

*Although Einstein and Rosen were probably the first to recognize the "Einstein-Rosen bridge," as it has now come to be known, the real insight into its structure came from the work of several physicists. These included Christian Fronsdal, G. Szerkes, Arthur Eddington (as far back as 1924), David Finkelstein, John Wheeler, Charles Misner, and most notably Martin Kruskal from Princeton University. I believe that it was Kruskal who made the issue of the bridge most apparent.

My Meeting with a Black Hole Map Maker

I met Martin Kruskal in the early 1960s when I was a graduate student at UCLA. My Ph.D. thesis work dealt with some of the mathematical complexities of plasmas—ionized gases that could be squeezed together in magnetic fields so that the ionized particles of the gas would bang into each other and produce thermonuclear reactions much like the way a hydrogen bomb works. Dr. Kruskal was an expert on such mathematical structures working with a group of fellow physicists at Princeton on the peaceful uses of controlled thermonuclear fusion, an enterprise called Project Matterhorn.

Being a graduate student in physics can be quite intimidating at times. With your environment of fellow "smarty-pants" graduate students and equally erudite professors, one can often feel the sting of inadequacy. My experiences were no exception. My usual picture of other physicists at that time was that they tended to speak mainly to those working on the specialties they were interested in and few others. A lowly graduate student wasn't usually worth talking to.

Even though I wasn't working on the same kind of problem as Martin Kruskal, I found him quite willing to talk with me, which in itself surprised me, and offer encouragement. Our meeting was quite brief (I doubt if he would remember it), yet I found him affable, willing to "open up" and share his remarkable insights into the mathematical structures of complex physics. I remember his asking me what I thought about a particular mathematical facet. He wasn't testing me, he was asking me as a fellow physicist, not as a student. I was quite taken by his remarkable mind and human sensitivity.

Kruskal was sensitive to more than human beings, however. He had realized that the Schwarzschild or critical radius of a black hole had the appearance of a *singularity.*

Singularities, you will remember, are those points in space or spacetime where physical quantities go berserk. Instead of varying normally as one moved from one place to another (or from one spacetime point to another), these quantities would become infinite in value as you mathematically approached them. Now approaching a singularity in mathematics simply means asking what happens when you look at the value of a physical quantity, such as a time interval, when you go near the region surrounding the singular point. If the point is singular, the values you are examining become infinite in value.

Mathematical infinities are one thing, but what about the real world? One has never observed such infinities, so physicists tend to regard singularities as evidence that their equations are in error. Kruskal and other physicists knew that when a test particle—a small particle with very tiny mass that is sensitive to the spacetime it flies in—reached the Schwarzschild radius of a black hole, time shrank to nothing and space stretched to forever. Thus many physicists regarded the Schwarzschild radius as a singularity not worth considering.

However, these same physicists saw that the singularity at the Schwarzschild radius did not have the same characteristics as a true singularity. They saw that this "spherical singularity" resembled the same kind of apparent singularity that was exhibited when using latitude and longitude lines to draw a triangle on the surface of a sphere when one of its corners was at the pole. Normally every triangle on a sphere had to have one leg drawn parallel to a line of latitude and another leg drawn parallel to a line of longitude.

A triangle could still be drawn, but it would no longer be the same because lines of latitude vanished at the pole. With a different coordinate map, using lines that crossed at a right angle on a plane, for example, the triangle could be drawn without any map problems. In a similar manner, the spherical singularity wasn't really indicating an error in physics but a poor choice in the form of the map used to picture the geometry.

Kruskal then created a new coordinate map surrounding the black hole which made the singularity at the black hole radius "go away." Now it would seem that Kruskal was performing as a kind of mathematical magician by doing such tricks as this. And in a way he was.

The idea was to find a new set of coordinates that also were perpendicular to each other but were not what we call space and time coordinates. It was like redrawing a map of Washington, D.C., but using only streets running north-south and east-west instead of the circles and radial lines presently used to lay out our capital.

However, these new coordinates would not exactly correspond to "someone's" measure of space and time, as the old coordinates did. In other words, Kruskal asked himself: Was there an observer that would see the space and time around a black hole in such a way that the black hole radius would no longer be singular? The answer was yes. However, the observer wasn't human. It was a zero-time particle—a speeding photon. From its point of view there was no singularity at the critical radius; however, something singular still remained to its imaginary eyes.

Now it might seem a little strange to speak of a zero-time particle acting as an observer of a black hole. What do I mean by that? First of all remember that a zero-time particle is a photon—a particle of light. So what I am asking you to do is to imagine what a black hole looks like to light itself. Einstein was always concerned with how the universe looked when viewed from a speeding photon's point of view. He imagined speeding along with a photon and looking at himself in a mirror. His question was: Could I see my own reflection since I would be moving at the same speed as the light striking the mirror? A little thought indicates that one would not be able to see one's face in the mirror since there is no relative speed between the light and the mirror. This bothered Einstein, and he came up with the theory of relativity which showed that he would be able to see his own face since the speed of light was the same for all observers even if one was moving at the speed of light itself.

Similarly Kruskal was concerned with the fact that light

slowed down to zero speed as one approached the edge of a black hole. It didn't seem reasonable for light—a zero-time particle—to slow down, for it wasn't moving in time or space as far as it was concerned. So he constructed a map in which the coordinate lines were lines actually followed by light as it made its way across the universe and into and out of the black hole. On this new map the black hole radius was no longer a singularity. Instead it became a pair of crossed lines—lines followed by photons as they circled around the black hole—caught by its gravitational field.

Even though the Schwarzschild singularity vanished on the new map when drawn with the new coordinates, another singularity remained. To understand this we need to consider that in the old coordinates there were two singularities, one at the Schwarzschild radius and one at the exact center of the hole, the "zero" singularity. There, the force of gravity became infinite. Crashing into the zero singularity was unavoidable once an object crossed the Schwarzschild radius threshold. This singularity was a real one and couldn't be avoided or mathematically transformed away. It was inevitable that such a crash would occur.*

On the old map, the singularity was at the center and appeared as a point. In the new Kruskal coordinates, this zero singularity appeared in two distinct regions of the new spacetime map. It lay in the future of all test particles making their way across the black hole boundary, and it also lay in the past. Thus the black hole had a more complex structure than was first suspected. And for the first time physicists realized that a black hole meant both the death and the birth of matter, simultaneously. A black hole was also a white hole —spewing out matter from the past singularity in just the reversed manner of particles all crashing into the future singularity and being gobbled up. The existence of two singular

*However, if you look ahead to Part Four, you will see that it was just this essential singularity that bothered Stephen Hawking. By bringing quantum physics into the black hole, Hawking even made this singularity disappear! But you had to add quantum mechanics to do it.

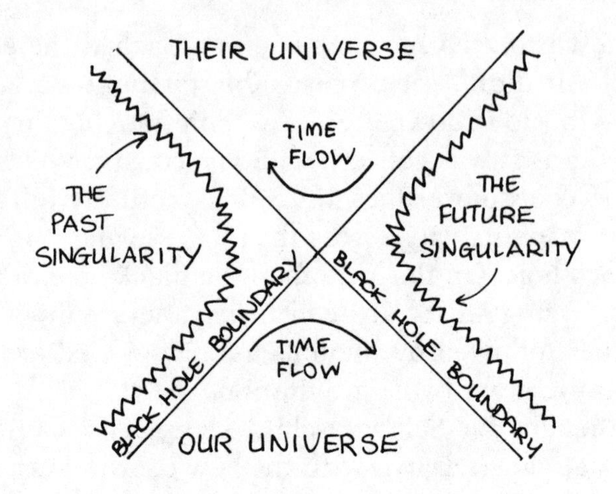

KRUSKAL'S BLACK HOLE MAP
SHOWING A PARALLEL UNIVERSE

regions inside the hole would seem to have gotten us nowhere in our attempts to rid us of singularities.

We got rid of one only to have another pop up. However, the new singularity gave the hole a symmetric structure, and to a physicist a new symmetry is a clue to a deeper understanding.

On the new map, four distinct zones could be made out. There were two outside universe zones and two inside universe zones (zones inside the Schwarzschild sphere). The two outside zones were regions corresponding to outside the black hole. One of these zones corresponded to our own universe. The other zone corresponded to another universe. Inside the black hole there were also two zones. A future zone and a past one. The future zone held the future singularity. Anyone passing through the surface of the black hole from our universe would inevitably crash into it.

However, just the reverse was true for the other outside universe. Anyone passing through the black hole boundary would crash into the past singularity. The other universe has (or will have) its time sense reversed from our own.

From the zero-time particle's point of view, space and time

were indeed one and the same. This was reflected in the coordinates chosen by Kruskal, for the pair of crossed lines that marked the four separate zones were simultaneously both lines demarking the critical or Schwarzschild radius and two fixed times—the beginning of time and the ending of time. Thus there was no singularity at the Schwarzschild radius of a black hole for photons. Instead they existed on the edges of eternity balanced between the two parallel universes.

Thus what was future for us was past for dwellers in the other universe. By having both a past and a future singularity, the black hole opened up into a bridge connecting two parallel universes. The problem was, could anyone cross that bridge?

What's Black and White and Dread All Over?

With the new map, a black hole became something quite "alive." It wasn't just a black ball floating out there in space, but a thing capable of distorting space and time to suit its own purposes, whatever those were.

In a joke, "What's black and white and 'red' all over?" the answer is a newspaper—a means for communicating information about one part of the world to another. The black hole had its own joke. It was "black and white and dread all over," because it contained these black and white singularities. It also was possible to find out about one universe from the other by traveling into the hole. But the trip may not be worth it.

It was both black and white, meaning that it could spew out matter (a white hole) as well as suck in matter (a black hole). In its white hole capacity, matter that happened to be born in the past singularity could escape. How? Nothing to it from its own spacetime perspective.

Coming from the past singularity, it would inevitably pass through the surface to our outside universe. Eventually it would be sucked up again by the future singularity after crossing the black hole surface for a second time. It was only for matter coming from the outside universe that the hole appeared black. All outside matter was gobbled up. All inside matter was belched out as if from the whale that swallowed Jonah.

On the other hand, matter that was crashing into the future singularity coming from our universe would appear to the parallel universe dwellers to be coming from their past. Since their time is time-reversed from our own, the future singularity is for them the past singularity. Our past singularity is for them a future singularity.

Thus they would see matter coming into the black hole from our universe appear as if it were being born and coming from the future singularity passing out toward us.

As one passes from one universe to another through a black hole, space and time undergo a reversal. Inside the black hole our time coordinates reverse, so that if we could watch a space traveler cross the black hole we would see him traveling backward in time as he approached the future singularity. The singularity would still be in the future for him, but not for us. In fact, if we could watch all this happen we would see two travelers, one approaching the black hole from the outside, and one escaping the black hole from the inside. The two would join up and pop together as one just when the traveler crossed through the Schwarzschild sphere.

To grasp this situation better, suppose we take an imaginary journey into a black hole, coming from our universe. Let's describe this journey not only from our own perspective as we move into the hole but also from the perspective of a viewer in our universe and from the perspective of a viewer in the other universe. This will clarify the white, black, and time-reversal characteristics of the black hole.

Chapter 16

AN IMAGINARY JOURNEY TO PARALLEL UNIVERSES THROUGH A BLACK HOLE

As I mentioned, a black hole is not a stagnant thing; it moves in time. In fact it drags time and space around with it much like a drain hole drags sewage down into itself. The hole can be pictured as a tube of rubber with two rims—one rim in our universe and the other in a parallel universe. The lines running down the tube correspond to the distance from the center of the hole. The lines forming circles around the throat of the tube correspond to different times. Thus a circle near the opening rim corresponds to early time, and a circle around the throat corresponds to a later time. As an object approaches the hole, the tube begins to stretch as if the two rims were being pulled apart.

Thus the distance to the center appears to be stretched away as the object approaches the center, and time appears to take longer. This causes the tube to squeeze together at its

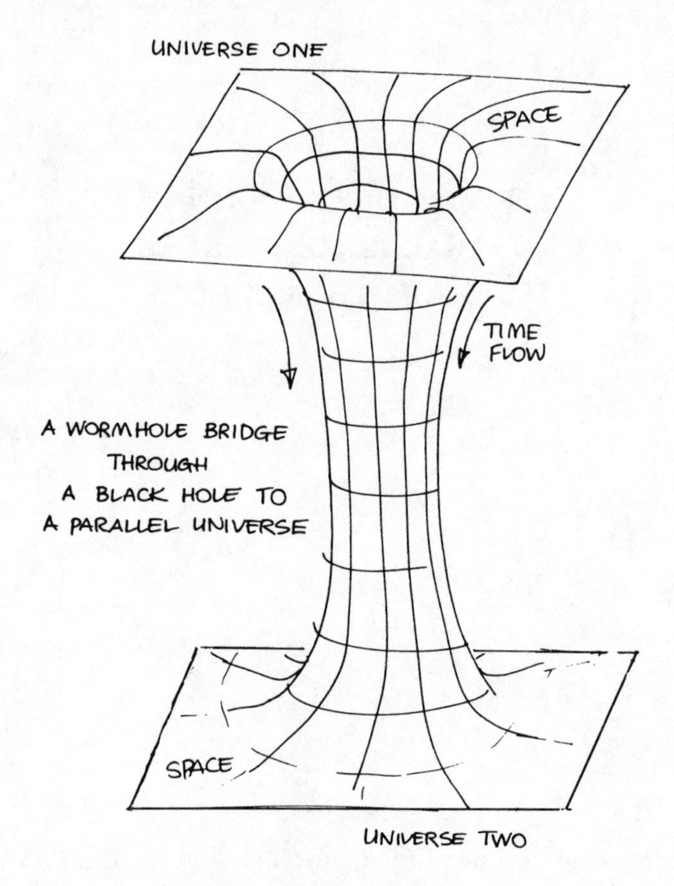

throat; and if the tube is pulled apart far enough, the throat closes off, trapping the object. This is the inevitable crash of the object into the future singularity.

If we now imagine a person traveling into the hole from our universe he would see quite an amazing sight. At first, as he approaches the hole, it will appear to be a black sphere in space. As he gets closer to the boundary—the Schwarzschild sphere's surface—the hole will still appear black and grow larger in front of him, much as any object you approach seems bigger as you get closer to it. However, space and time are also being dragged into the hole.

Consequently light that is coming from in back of him and circling around the hole, he will see as light appearing to come from in front of him. If he were able to stop his movement he wouldn't see this dragged light.

It is because he is approaching the speed of light that this distortion of light coming from his rear appears to him as light coming from in front of him. Also he won't be able to stop his progress; he is becoming a part of the dragged space and time surrounding him.

As soon as he crosses the critical radius, he will begin to see a point of light at the center of the hole. Darkness will surround the point of light, and a halo of light from his universe will surround the dark. Thus he will see a dark sphere with a light halo and a point of light at the sphere's center. As he gets closer to the singularity at the center, the point of light will bloom out into a sphere of light.

This light will be coming from the parallel universe. Light from our universe will continue to halo around the now decreasing blackness, making the sphere appear as a black rim. The rim will decrease in thickness as he moves closer and closer to the center. Just before he crashes, he will see mostly light, with the rim of blackness nearly vanishing. Thus he will witness events in our universe and events in the parallel universe.

However, what he sees will be fantastic. The light coming from our universe, as he first approaches the hole, will be light coming from a universe that appears to be speeded up. It would be like watching a movie with the number of frames per second approaching the millions. He will see our whole universe die as all the stars go out; or perhaps it will die, not in a whimper, but in a big crunch, with all matter being swallowed up in some gigantic black hole appearing elsewhere.

Just as he crosses the critical Schwarzschild surface, he will see infinity. All of the universe's history to be will pass before him in a flash—for the universe he left will have aged remarkably quickly.

He will then cross over into the black hole and begin to see the other universe. However, he will still see a light halo coming from his universe. This halo will appear to him as the same movie he just watched, only this time running backward in time. Thus he will see all the events of his past coming to him in a time-reversed sense.

He will first see the big crunch or the stars go out but since this is time-reversed, actually this will appear to him as a big bang or the stars suddenly popping on.

As time passes backward he will watch all the events surrounding his coming into the hole. He will see himself receding back toward the launching pad on earth. Just then he will stop seeing anything more from our universe as he crashes into the very center of the hole where the laws of physics no longer apply. This is the dreaded singularity.

However, when he crosses the critical threshold into the hole he will see light from the parallel universe. This may also appear as time reversed. Thus if he observes a big bang in that universe he will be seeing a big crunch in that universe's future. He will see events from their universe also running backward in time.

For viewers at home in our universe his trip will also look weird. Since time is ordinary time for them, his journey will look as if he were slowing down as he approached the hole. He will appear to go slower and slower, taking forever to just cross over.

However, if he could somehow send light out to them when he went inside the sphere, they would also see him moving backward in time as he approached the singularity.

The problem is that light will not escape from the black hole. It will be squeezed off as it tries to make its way out of the hole. Thus if he could send messages that moved faster than the drain swallowed them, he could let us home viewers witness his crash.

Sending Tachyons to Do a Photon's Work

But to do this he would need to send tachyons, particles that go faster than light. Tachyons will not be bothered by the squeeze-off.

They can escape the squeeze and pass out of a black hole even though they were emitted by an object entering the black hole and crashing into the singularity. In fact, tachyons will pass freely from one universe into the other, avoiding the future and the past singularities altogether.

We can imagine that there are people living at the other end of the black hole. People just like us watching the spectacular display in their night sky. What will denizens in the parallel universe witness? Well, to them the black hole is also just a sphere in space. If they don't go close to it, they will not see anything happening. But if they also send a traveler into the hole, she will crash into the past singularity as inevitably as our traveler crashes into the future singularity.

Remember what's future for us is past for them and since their universe runs time-reversed from ours, their future is also our past.

Could the two travelers ever communicate with each other? The answer depends on whether or not they are really living in reversed time streams. If their future is our past, not only will they not see each other, but something I explained about seeing their universe as our traveler passed into the hole is changed. Their light will travel forward in time in their universe and not ever be seen by our traveler as he crosses into the twilight zone. All of their light will pass into the past singularity. It's going the wrong direction in time, so to speak, to be seen by him.

Thus she will not see him and he will not see her. If, by the way, she does come from a universe running backward in time compared with ours, she will also be made of antimatter—but that's another matter to worry about.

But even though her past may be our future as far as we can tell, we could be wrong. What's going backward for us may indeed also be backward for them. Their universe, in other words, may actually proceed normally from what we calculate is the past to the future.

They may experience all the normal things that we do, but be running in the same time direction as we do. What we call ahead in time or plus five minutes will be what they call plus five minutes, but would register in our calculations as time going backward five minutes. In other words, their universe runs the same as ours, but our time maps are reversed, so that we call +5 minutes, they call −5 minutes, and vice versa.

Now as weird as this appears, it's no more than a matching of our plus time with their minus time in just the same manner as you might match yourself with your mirror reflection. Time runs in the same direction in both universes, but we calculate their time in a reversed sense. They would say a similar thing about our time.

In this case, the future singularity for us would also be the same future singularity for them. Then the two travelers would be able to see each other. She could watch our universe and he could watch hers as they crashed together into the future singularity. Each would see their own universe running backward in time but see the other universe running forward in time. Oh, well, as Saint Augustine once remarked, "If you ask me what time is, I don't know the answer. If you don't ask me, I do."

Now I mentioned that a journey out of a black hole was possible. For this to happen a traveler would need to begin the journey from the black hole's interior. He or she could not start outside and then cross the surface twice. For this reason the surface of a black hole is also called an *event horizon*. An event horizon is the surface of a sphere marking the edge of a black hole. It is called a horizon because like a horizon at sunset you can only approach it but never actually reach it. It takes an infinite amount of time to approach the event horizon, as measured by observers watching the spectacle from a distant viewing spot. However, only a finite amount of time is observed by the person approaching it. Again relativity of time shows its head. If you do manage to cross the event horizon and enter into the interior of the black hole, you can never turn back and reach the event horizon again. You will be swept away by the flow of spacetime inside the hole, eventually ending your existence at the black hole's center. The idea is that you can't cross an event horizon twice—one event horizon per universe experience is allowed, no more.

How a Rotating Black Hole Is a Bridge to Many Universes

The above statement about never crossing an event horizon twice is true, provided there is only one event horizon. Thus the statement should be, "You can't cross the same event horizon twice"—much as you can't enter the same river twice. The reason is that the event horizon marks a boundary between different spacetime orientations.

In our universe, we think of ourselves as sweeping through time without recourse, but moving through space both backward and forward. But once you cross an event horizon, time and space reverse. Space becomes a "stream" that has no recourse and time becomes "spacelike," allowing you to move back and forth freely. What this means is that once you are over the event boundary you are swept forward in space as perceived by the outside universe. The space inside the horizon has a "timelike" characteristic, meaning that you must go with the flow, while the time has a "spacelike" characteristic, meaning that you can flow in either direction. Thus even though you, the traveler, experience a steady movement in time as you perceive it, those mapping your motion will see you move in time as you enter the event horizon but then you will be perceived to sweep through space toward the hole's singularity while going back in time. If the outside world could watch you while you were doing this, it would appear that you were going backward through time as they experience time.

Thus crossing the event horizon causes a reversal. What was time in our universe becomes space inside the black hole. Eventually you will run into the singularity.

Now all of this is true for a hole that is not spinning. It is remarkable that a black hole could be spinning as well as having all of the other weird properties I mentioned.

How could a black hole have spin? Physicists have noticed most stellar objects are spinning on an axis. Our earth is no exception to this rule. Our sun spins on a regular cycle. So do Saturn, Jupiter, and our own moon. Most stars spin. If any of them are overweight, they too will collapse into black holes and maintain their spins as they do so.

And a spinning black hole has another weird characteristic—it has two event horizons, an outer and an inner. The outer event horizon is the same as the event horizon for a nonspinning black hole, but the inner event horizon is a reversal of the outer. Thus when you cross over the inner event horizon, it's like passing into the eye of a hurricane—time and space become "normal" once again. You can "play" inside the inner event horizon and need not crash into the singularity.*

Just as Martin Kruskal gave us a new map of a black hole, in 1961, and showed us that a nonspinning hole was a bridge to another parallel universe, Australian mathematical physicist Roy P. Kerr, in 1963, then at the University of Texas, discovered a complete solution[6] for the rotating black hole.

Needless to say, Kerr's solution was marked as one of the most important developments in theoretical astrophysics of the mid-twentieth century. The factor we want to examine is that Kerr's equations indicated the existence of an infinite

*The singularity is weird, too. There is a negative space inside of it where gravity reverses and becomes repellent instead of attractive. If you happened to enter the negative space, you would be spit out like a disgusting meal to a vegetarian giant.

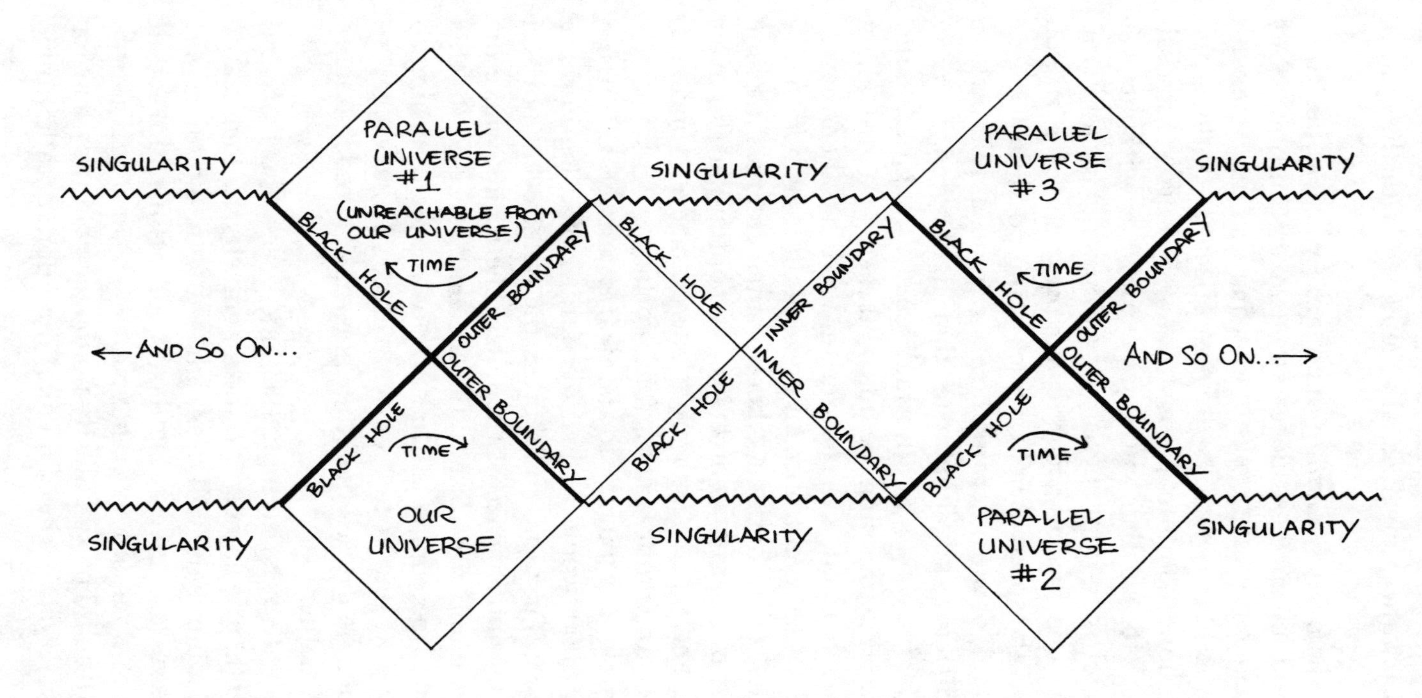

THE INFINITE PATCHWORK OF PARALLEL UNIVERSES IN A SPINNING BLACK HOLE FOUND BY ROY KERR

number of parallel universes, all connected with the spin of the hole.

Because the hole is spinning, there are two event horizons, an outer and an inner. A spacetime voyager can now safely pass from our universe into any other parallel universe with the exception of one parallel universe. Because of the structure of the honeycomb map constructed by Kerr, it is possible to travel to all other universes except one without ever moving faster than light speed. The exceptional parallel universe is the one just adjacent to our own. To reach it, you must exceed the speed of light, which is not possible.

However, all of the other universes are reachable. Nowhere does his trajectory take him faster than light speed. To make a journey he must pass at a normal speed through the outer event horizon surrounding the hole and then pass through an inner event horizon. Continuing onward he then passes through an inner event horizon once again and then finally another outer event horizon, where he then emerges into one of two possible parallel universes. By continuing this way, weaving in and out through inner and outer event horizons, he can reach any number of parallel worlds.

For example, his journey to the nearest universe, starting in the year 2000, would be just like falling off a log into the spinning hole. First, he would cross the outer event horizon and enter the inner space of the hole—the region where his normal space and time orientation become reversed. In this region he will be compelled to move on, never returning to our universe again. In fact our universe will have aged to infinity when he crosses the outer horizon. There is no more universe back there!

Next he will pass through the inner event horizon and enter the zone just adjacent to the singularity. As long as the hole keeps spinning he will pass right on by the singularity. It is no longer a "spacelike" singularity compelling him to fall into it. Since time and space have reversed once again, time is back on track. The singularity no longer holds any fear for him. It is oriented in the "timelike" direction, parallel to his movement in the hole. Of course, he must be careful to pass by the singularity and not get detoured into it.

Soon he will pass out of the innermost region, passing once again through an inner event horizon. Again time and space will reverse for him, and he will be compelled to move on out. When he crosses the outer event horizon, he will then pass into a parallel universe—a whole new world. In fact, he will witness the birth of that universe as he crosses the last horizon.

If his trajectory is true enough he will pass into his parallel past, possibly reaching the time of his birth. If the second universe is an exact copy of his universe, he will see himself being born again.

An Infinity of Parallel Universes

Not only are there four parallel universes connected through a spinning black hole, but there are an infinity of them. Kerr's solutions have shown that the map extends on forever, both toward the past and toward the future. Could these universes ever communicate with each other? Could there ever be any physical phenomenon observable in our universe that would show this connection? Aside from actually making such a journey as our fictitious traveler has done, there is another exciting possibility. Perhaps the infinity of parallel universes predicted by Einstein's general theory of relativity as shown by Kerr are the same parallel universes predicted by an entirely different theory of physics—the quantum theory. If this is so, there will be a number of exciting developments. We will look at them in Parts Five and Six.

In Part Four we will explore how parallel universes were predicted to occur at the very beginning of time. As we saw in Part One, quantum physics dealt with parallel universes as if they were otherworldly ghosts of probability that could intersect with our world. In Part Three we realized that relativity posited their existence through regions of spacetime that were highly distorted. Perhaps these quantum physical probability ghosts are images coming from such spacetime distortions. Perhaps they are signals of the existence of the same universes predicted when time and space began.

To explore these possibilities we need to look at cosmology —our theory of the beginning of the universe. With the new light shed by relativity theory we know that space and time are quite different than conceptually envisioned by Newtonian science. With quantum theory we know that matter is also very different than prequantum physicists could imagine. Matter's properties depend on choices made by observers of matter. These properties must coexist as possibilities— each in a separate universe, according to the parallel universes interpretation of the mysterious quantum. Since relativity theory predicts parallel universes arising from distortion of spacetime, and quantum theory predicts them because of the coalescence of possibilities, it now appears sensible to regard these universes as the same. This promises to bridge the gap between relativity theory and quantum physics if physicists can do it in a consistent way.

However, any attempt to bring quantum physics into the fold must also solve the problem of observation, particularly at the beginning of time. How did the universe spring into existence if there wasn't any observer, then? If there was an observer, what role did he or she play? In Part Four we will explore the role of quantum physics at the beginning of time with and without *The Observer* and how how parallel universes offer a consistent theory about how we all got started.

PART FOUR

"IN THE BEGINNING THERE WAS . . ."

In the beginning the universe was created. This has made a lot of people very angry and been widely regarded as a bad move.

Douglas Adams
The Restaurant at the End of the Universe

Cosmology deals with the theory of the early universe, how all that we can imagine as physical—matter and energy—first began about 15 billion years ago. This theory has gone through a number of important changes. We now realize that earlier theories of cosmology, even those that have included Einstein's theory of relativity, must still be wrong because they fail to include quantum physics in their deliberations. By including quantum physics, we find strong and surprising evidence for the existence of *parallel universes* at the very beginning of time.

In Part Four, we will explore the ideas of *cosmology* including the ideas of quantum physics. As we will see, cosmology —usually taken to mean the *big bang,* or sudden explosion at the beginning of time—as a theory without quantum physics has several inconsistencies.

These inconsistencies arise because cosmologists are attempting to fit Einstein's general theory of relativity together with Newtonian physics leaving out anything to do with quantum physics. The whole question of how properties of matter arise as choices made by an observer is not answered if one leaves out quantum physics. If the universe arose in a specific energy state, then someone had to choose to observe the universe in an energy state. But if the universe is all that there is, where and when did that observer occur? If the observer is part of the universe itself, then how did the universe give birth to itself? The only hope of reconciling these and other questions must include quantum considerations. There is no hope of grasping the universe based solely on Einsteinian relativity and Newtonian mechanics.

Cosmology is based on a plain classical Newtonian fantasy. This fantasy assumes that all matter behaves as if it were made of tiny pointlike particles that move through infinite space and time with predictable positions and speeds at each and every moment. This fantasy, although quite useful when dealing with large amounts of congealed matter, such as baseballs and automobiles rolling through the streets, doesn't hold water when dealing with atomic and subatomic particles, particularly when they are being created at the time of the big bang. Something more is required.

Nevertheless, the early theories of the big bang were based on classical physical concepts involving the theories of ordinary *gas dynamics* and *thermodynamics*. According to these theories, the universe began as a point that exploded into an expanding fireball. It is clear today that this early theory is inconsistent.

Most cosmologists now agree that there cannot be a consistent theory of the big bang or any early scenario without bringing together all of the present knowledge of physics. Later attempts have certainly taken into account relativity theory. However, although relativity has increased our knowledge of the big bang and in itself points to a new vision of space, time, and matter, relativity is a classical theory as well. Since classical physics has long ago been replaced by quantum physics to explain the present universe, it stands to reason that the concepts of quantum physics must play a role in the big bang.

But this turns out to be fraught with controversy. No one quite knows how to do it. The major reason for this is that quantum physics has at its heart the important role played by the actions of observation. These actions alter physical matter. The properties of matter that we observe every day, such as the hardness of metals and stones, the temperatures of heated gases, the color of light and materials, as well as the refined properties of atoms and molecules, depend on what is chosen to be observed. If matter remains unobserved, quantum physics predicts that matter will not have any of these properties. It will exist only in the form of probability patterns—patterns that enable these properties to be observed.

These patterns of probability are quite weird. They are themselves unobservable. But whenever an observation of matter occurs, these patterns suddenly change, with the result that matter appears with the property sought for. Not only that, but these patterns can act together producing a new physical possibility—much as superimposing a number of transparent images in a slide projector can produce an image that is not contained on any of the individual transparencies. Synthetic fabrics are only a tiny example from today's world of quantum technology where the superposition of possibilities to produce a new possibility in the physical world is used.

This new possibility cannot occur unless the separate possibilities act simultaneously. To carry the slide image analogy further, assuming that each slide carries a particular property of matter, it is as if the properties of matter depend on the choices of the exhibitor of the slides. Thus all of the observable properties of real matter needed to explain the creation of the universe can exist only through the superposing of these separate possibilities.

Each possibility must be distinct from all others, since each involves the appearance of matter in the physical world of space and time with a particular property. Each possibility must occupy a separate region of space and exist for a particular interval of time. Somehow these regions and intervals, the arenas in which matter plays the game of physics, must be adjacent to each other. Yet each possibility must also occupy the same space simultaneously in order that these patterns of probability can be superimposed together. Thus the arenas must all be the same arena (matter must occupy the same space, our universe) and, at the same time, an infinite number of adjacent arenas (a number of spaces or universes that are somehow "disincarnated" from each other).

Such *spaces* are today called *parallel universes.*

Consequently, whether or not the big bang occurred before there were observers, quantum physics states that parallel bangs had to occur—each bang in a separate but equal other universe.

The First Inconsistency, Not Enough Time to Start the Universe

According to what has now come to be known as the "standard model" of cosmology, *the* universe—only one—began with a finite but quite small radius (about the size of a human blood cell). Carrying on from this point of view the first problem occurs. The universe, as small as it was, was still too big to begin with. Not enough time was present to allow light to reach all the corners of the early blood-cell-sized universe. As the universe expanded, this problem persisted up to and including the present day.

Thus, our present universe is still too big to reconcile with this big bang model. There isn't enough time to balance the temperatures in the whole universe. Therefore, regions of the universe vastly separated from each other should show different temperatures. But they don't.

Inflationary Models

To create a balanced condition, some physicists have attempted to construct a model of the early universe based on the idea that what began, began from almost nothing at all (much, much tinier than a human blood cell) and then went through an *inflationary phase* in which the whole universe expanded faster than light.

The reason for this inflationary theory was to remove any problems associated with all of the infinitely many possible starting conditions that could have arisen when the universe began and, at the same time, provide a mechanism for establishing equilibrium. By starting the universe in a much smaller region so that light could travel across it at the beginning of time, there would be enough time for light to pass on the required energy from one place to another before the inflation took over.

This earlier tiny model coupled with the rapid inflationary phase would enable the universe to expand already in equilibrium and would wipe out any initial-condition anomalies, much as a rapidly expanding balloon wipes out any wrinkles in its deflated skin. In other words, any number of starting conditions would be able to reach the same final goal—our universe today.

Inflation Is Not Enough

But this, too, leads to inconsistencies if the universe were alone—the only one to bubble into being. The major problem is that still not all starting conditions would be wiped out (we should see some evidence of them today) with inflation, and, even if they were, such a model still does not operate according to the laws of quantum physics.

Including quantum physics in the beginning of time introduces all of the quantum weirdness that presently inhabits the scientific world today. Particularly, since no *observer* was assumed present at the beginning of time to choose the observable properties of matter, one must include the possibility that all possible parallel universes also appeared when our own universe appeared.

With or without the observer, these other universes are required to stabilize themselves in much the same manner that bubbles arising in a liquid form a stable bubble raft of foam on the surface of that liquid. Such stability is related to the stability of an ordinary atom as determined by quantum mechanics.

An atom is composed of subatomic particles called *electrons*. One must have, in the quantum picture, many possible parallel locations for each electron in an atom in order that the atom exist in a stable pattern of lowest energy. Each location is a possibility for each electron. In this lowest energy state, called the *ground state,* the atom has a well-defined minimum energy, but its electrons do not have well-defined locations. To maintain stability in an atom, electrons must not have well-defined locations. Thus the atom can exist only if its electrons are in ghostly parallel-location worlds—each electron somehow occupying only one position in any world, but occupying an infinite number of positions in an infinite number of nearby parallel worlds, *all at the same time.* In a sense, this can be visualized as a raft of parallel electrons floating in a sea of possibilities.

However, the analogy goes much farther. These additional bubbles not only do not exist in just one universe, they exist in parallel universes—universes that can be reached from our universe via a process called *quantum tunneling.* In this process an electron is able to suddenly vanish in one universe and appear in another. Indeed, if this idea is correct, much of what we now call psychic phenomena, altered states of awareness, channeling of conscious beings, spirits, ghostly apparitions, flying saucers, and other unexplainable phenomena could be explained as information tunneling—coming from parallel universes.

Parallel Universes Solve Another Problem

The bubbling into existence of parallel universes also solves another problem. That is the *why us* problem. Why are we so fortunate to be in the only universe that is? Just alter the early conditions of the universe by a hair of probability, and we vanish in a puff of cosmic smoke. By including parallel universes we include those universes where life as we experience it is not possible as well. This provides a solution of the *why us* problem because it enables all possible universes to exist side-by-side, even those that do not have just the right starting conditions to engender life as we know it.

In Part Four we explore these questions about the beginning of all that is and show why we must have parallel universes at the dawn of time.

Chapter
17

THE FIRST OBSERVER OF THE BIG BANG

For some reason we *Homo sapiens* like to wonder about how things began. If we don't know (and even today we don't know), we imagine. Our musings about how it all got started depend on when we are living. Today we use the language and rationalization of modern science to tell the early story. Our earliest ancestors, so it is written, imagined all kinds of start-ups. As Steven Weinberg, physicist and Nobel laureate, explained in his book *The First Three Minutes,*[1] our early Norse ancestors imagined, as chronicled by Icelandic magnate Snorri Sturleson in the year 1220, that in the very beginning there was nothing. No heaven, no earth. Nothing existed except for a huge gap of space. There wasn't even a blade of grass. But, and here the story has a few more gaps, to the north and south of nothing there did exist some regions of frost and fire. The heat from the fire licked across the big gap and melted some of the frost, out of which there grew a giant and a cow and then some grass for the cow to eat.

From our vantage point of time such stories seem simple and filled with holes. However, we shouldn't be too hasty in our glancing down our nose-sides. The early Norse were certainly correct about one thing, as far as we know; in the beginning there probably was nothing, not even a blade of grass.

The Names of God

The Bible informs us that in the start-up condition there was, first of all, God. And then there was this division of waters. And at some point God let there be light and there was light. Farther on we read that God began to name His manifestations. Each act of naming a thing, thus, in a sense, created that thing as a reality. God's role was, then, that of both the creator and the observer of all things. It would appear, therefore, that perhaps creation and observation (wherein a thing is seen as a thing and called a name) are equally important at the beginning of time.

This idea turns out to be important in quantum physics, something I have called in my earlier books[2] *the observer effect;* and it is also important in attempting to describe the quantum physics of the early universe and the role that parallel universes played when things began.

The *observer effect* is the sudden change in the probability of observing some property of matter, such as its location in space, when that observation is actually carried out. If no observation actually takes place, the thing remains without possession of the tangible property sought for, in this case a location in space. It is said to be in a probability or quantum wave pattern that I call a *qwiff.*

The well-known wave-particle duality of quantum physics is an example illustrating the observer effect. If the observer chooses to observe the wave nature of matter, he or she gives up or has no hope of observing the particlelike nature of matter, and vice versa. Thus, what is observable depends on the choice made by the observer. In this sense the observer has much input into the physical properties of the matter observed.

A Cloud and the Observer Effect

As a qwiff—a probability pattern distributed in space and throughout time—the thing can be thought of as existing in all of its probable locations simultaneously, like a cloud of water vapor surrounding a central point. Each invisible droplet represents a possible location for any visible drop.

To observe a drop, a long straw is imagined to be inserted into the cloud. If an observation occurs, the straw disturbs the cloud and causes the cloud to momentarily condense, forming water, which is sucked up the straw as a tiny drop. If an observation does not occur, the straw does not disturb the cloud. No water is drawn.

Just where a drop will condense in the cloud depends on how thick the cloud is. If the cloud is uniformly thick, any point is equally likely to produce a drop. But when the drop is sucked up in the straw, the cloud vanishes. Now the remaining points where the cloud previously existed are void of the possibility of producing a drop of water.

The Observer Disturbs and Creates

Thus, the observer effect not only changes the odds, it brings into existence the thing being observed. In essence, to observe a thing is to create the thing being observed.

The question is, who was around to observe the early universe? If there was no observer to begin with, that is, God did not watch creation, then according to quantum physics all possible results of any observation must have arisen side-by-side at the same time. This is what the *parallel universe* theory of quantum physics predicts.

In a slight variation of this, one pictures an observer to be nothing more than a recording instrument. When an observation occurs, say, the recording of a tree falling in a forest, the recording and the tree it is observing enter together into parallel universes—each corresponding to one of the possible locations of the fallen tree and the recording device containing that particular record.

In this version, God, the first observer, is caught, like an unfortunate Br'er Rabbit by a Tar Baby, by merely observing His or Her own creation. Perhaps that's the answer—we are somehow God caught in the morass of materiality, because we wanted something to look at.

Before we look into this, let's consider some of the problems associated with the beginning of time, without quantum physics or parallel universes.

The Grand Prix at the Beginning of Time

Today, the most accepted start-up is reminiscent of a Grand Prix. In the beginning, gentlemen—cosmic race-car drivers—started their engines and there were explosions. In terms of classical physics, there was only one explosion at the very start, called in the tradition of Dashiell Hammett *the big bang.*

Time at the Beginning of Time

This explosion didn't start just somewhere in the cosmic car, but everywhere all at once. However, everywhere all at once doesn't mean the same as we might try to imagine—for at this epoch, or should I say instant, as instants and epochs are the same thing during the early universe, time was also just beginning. And believing Einstein for a moment, since time and space are aspects of one continuum, space was also just beginning.

Space Where No One Has Gone Before

Thus one is led to the idea that everything started at one tiny point smaller than a pinprick. The best theory today states that the early explosion started in a volume smaller than the nucleus of a hydrogen atom. To imagine how small this space is, remember that one atom of hydrogen compares in size to your thumb as your thumb compares to the whole planet earth. The nucleus of the hydrogen atom, called a proton, compares in size to the whole atom as a Ping-Pong ball compares in size to a modern football stadium.

Timewise it is even more difficult to imagine the first instant. Compared with one second the first instant was extremely quick. Today we commonly think in terms of nanoseconds—billionths of a second. In one nanosecond light travels a little over one foot. One nanosecond is to one second as one second is to about 32 years. Suppose we consider an even smaller time interval—one attosecond. One attosecond compares with a nanosecond as a nanosecond compares with a second. Thus one attosecond compares with a second as a second compares with about 32 billion years. Our present universe is only 15 billion years old.

However, even one attosecond is long compared with the first instant. That first instant compares with one attosecond as one second compares to 32 billion years. In other words the first instant compares to a second as a second compares to 32 billion billion billion years. On a scale of instants, the universe grew very, very old before the first second of time had elapsed.

Now this was the way it was according to classical physics. All space was tinier than a proton and all time was less than this first instant. And all matter was contained in this all-space during this all-time period.

It is obviously difficult to imagine such a thing as this. According to Einsteinian thought, space, time, and matter are quite dependent on each other. As Einstein once put it in a newspaper interview (and I paraphrase), "Before relativity we thought that if all the matter in the universe were to suddenly vanish, space and time would still exist. After relativity we now think that they too would vanish." Thus like the horse and buggy, each defining the other, space, time, and matter each provides the background for the definition of the others. Space without time cannot exist because of the fixed speed of light traveling in a vacuum and matter without both makes no sense at all. Einstein showed that matter affects spacetime, causing it to "curve" in on itself. And surprisingly, when matter is sufficiently dense, this tight curvature turns out to provide a passageway or tunnel between one parallel universe to another.

Black Holes: Another Way to a Parallel Universe

Such a passageway exists, according to the theory of general relativity, inside of objects that we now call *black holes.* We already saw how relativity predicts other universes, black holes, and their hidden passages in Part Three. Here it is useful to grasp the scale upon which a black hole exists, and also the scale upon which the universe began.

The Scale of the Universe

Think of a ball bearing, about one centimeter (half-inch) in diameter, filled with the heaviest material around, weighing two-thirds of a pound, give or take a few ounces. Even at this weight it affects the spacetime around it very little. But now imagine that we somehow manage to squeeze the earth down to this size.

In fact, now imagine that we squeeze around three hundred such "earth-balls," each about one centimeter in diameter, inside a Ping-Pong ball. This Ping-Pong ball weighs the same, has the same amount of mass, as three hundred earths. This ball is dense enough to form a *black hole.*

Its density is so high that light is bent just passing by it. Its gravitational pull on us is equally incredible. Each of us would feel a force 300 times our present weight just standing near it. In fact this dense ball distorts time and space so much that matter will be pulled into it and literally disappear from our universe, entering another parallel universe.

But even this extreme situation is far less bizarre than the early universe. To gain the conditions of the universe just at the time of the first instant, imagine squeezing one trillion trillion trillion times more mass inside the same Ping-Pong ball. This is probably beyond our imagination. But if we could see this, we would be witnessing the beginning of time as told by the classical cosmological mythologists.

At this moment all this matter fitting inside such a small space would be extremely hot. So hot that matter would be broken down into the simplest units possible, called photons, or particles of light. Such a universe is indeed simple. It is a gas of light.

Chapter 18

PROBLEMS IN EDEN

But even as strange as the first second of time is the stage of the universe before that. It is possible to extrapolate even farther back in time than one second. It is back then that parallel universes really are born.

However, it is not clear what we mean by the birth time, since time itself is part of the woof and weave of space-time-matter during this phase. This time is the "time period" beginning with the first calculable time scale. It is the time known as the first *chronon.* This time is extremely small when compared with the time just described. It is one billionth of the first instant. In terms of the first chronon the universe is quite old by the first instant. If we call the first instant a "standard," then one billion chronons have passed to make up one standard.

Now to grasp this scale of things, remember that one billion seconds is just under 32 years. Thus in terms of chronons, something equivalent to 32 years has passed before the first light emerged from the first point of space and time.

Here we are looking at the very beginning of the beginning as far as the laws of physics can go. And here is where we need to reconsider the whole question of the start of all things.

If we use our classically conditioned minds and just extrapolate back in time as I have been leading you, we find, of course, a single point with no diameter. Going back in time is like imagining a spherical balloon getting smaller and smaller each instant. As the balloon gets tinier, shrinking down to atomic size, down to nuclear size, and even tinier than that, its round surface curves more and more in on itself. We say that its curvature is ever-increasing as its radius is ever-shrinking.

Finally, when it reaches zero radius, it has infinite curvature. Such a region of space is called in mathematics a *singularity.*

Singularities Do Not Admit Laws of Physics in Their Domains

Now the problem with a singularity is, I admit, a philosophical one, for singularities will not admit any laws of physics into their ever-shriveling boundaries. We physicists must simply raise our voices in feeble protest as we ponder their existence. We are not allowed inside the proceedings.

This, of course, bothers some physicists, who believe that at no time or place would God allow a totally lawless universe to exist. Cosmological physicist Stephen Hawking, from Cambridge University, England, shares this sentiment.

Hawking is the author of *A Brief History of Time,** and is considered by most physicists as one of the most brilliant minds of the present century. He also suffers, as is probably well-known by now, from the debilitating disease known in this country as "Lou Gehrig's disease." This keeps him confined to a wheelchair and has affected his vocal chords and other body movements so that he needs to use a computer with an artificial voice channel to communicate his ideas to his assistants and students. He apologizes for what he calls "his American accent" because the computer voice was constructed in the United States.

Hawking has been attempting to merge quantum physics and general relativity theory together. In particular, he has been working at developing equations that would tell us just what did happen when time began. His theories use such concepts as imaginary time (see Part Three), singularities, and parallel universes to unite relativity and quantum physics. Hawking provided us with the idea that black holes really exist "out there" and attempted to give us theoretical means through which experimenters observing the distant galaxies could find them. If a black hole is detected, there is no doubt that he will receive the Nobel prize for his insights. He may get one anyway. At a 1987 workshop on quantum cosmology—a study group looking into incorporating the laws of quantum physics into the beginning of time—held at Fermilab in Batavia, Illinois,[3] Hawking reminded the participants that singularities have been tolerated for over twenty years, particularly the big singularity called the big bang. He brought into consideration the fact that there is no reason why only one singularity would arise. If one looks carefully enough at a singularity, one must realize that our universe must contain many of them, perhaps even an infinite number.

*Hawking, Stephen. *A Brief History of Time.* New York: Bantam Publishing Company, 1988.

The reason for this lies in the nature of quantum physics when applied down at the level of time of the first chronon and the volume of space occupied by the first speck of existence. At this level of reality, space, time, and matter are constantly fluctuating. It would be analogous to looking at a finely woven fabric in various stages of inspection. At first the fabric appears smooth and unfurled. But as we look deeper into its structure we begin to see the threads unraveling, and even deeper we see the spaces in between the threads.

At the smallest level of space-time-matter, spacetime is continually fluctuating—creating momentary bubbles of matter, which just as quickly vanish into nothingness again. Such bubbles do not appear only at one place, they bubble everywhere as a kind of frothy quantum foam.

Similarly if there was a big bang, there had to have been an infinite number of them, bubbling into existence more or less at the same time—if we can even think about what that would mean, since time only begins for each bubble when the bubble manifests.

Hawking has been working on the physics surrounding singularities for over twenty years. And he is not satisfied with the results. He stated at the meeting, "We cannot predict what comes out of a singularity.... It is disaster for science."

Because of this, and his faith in the lawfulness of the universe, Hawking now believes that there never were any singularities in the universe, not now and not when the Garden of Eden sprang into reality over 15 billion years ago.

How can we get around these singularities, when not only cosmology says they are present, but also general relativity? Hawking proposes that at all times the universe followed laws of physics, and that means that even at the beginning of time there had to be *quantum mechanics*, and that means *parallel universes* before the first observation occurred.

In the Beginning There Was Uncertainty

If we bring quantum physics into the first folds of time, there would not be any singularities. This fact has to do with the connection of momentum* and location that must exist for every object, called the *uncertainty principle (UP).* Accordingly, any time you try to squeeze an object into a space that is too small—in other words, locate the object—it will resist the squeeze by making its momentum more and more uncertain. This leads to its eventual possession of a large but undefined momentum—large enough to spring it from the trap of confinement set for it.

In an ordinary atom of hydrogen, which consists of a single nucleus and a single electron, the nucleus exerts a confining electrical pull on the electron in its vicinity. Without quantum physics, that electron would vanish into the nucleus—pulled irresistibly by that electrical force. But the closer it becomes confined to the nucleus, the greater is its uncertainty in momentum. And thus it can never be gobbled up, for it will quickly move away.

According to quantum physics, as I pointed out in Part One, the electron, instead of being pulled into the nucleus, takes on the appearance of a spherical cloud surrounding the nucleus. Like the water cloud mentioned in the previous chapter, the electron splits into multiple copies of itself, no copy actually being under the force that holds it, but all copies dealing with the confinement by forming a ghostly cloud.

The most probable radial separation for the electron from

*Momentum is a measure of matter in motion. A large hunk of matter moving slowly has large momentum because of its mass. A small bit of matter moving quickly has a large momentum because of its speed. It turns out that in quantum physics, momentum is a primary quality. Thus it is possible for an object to have a well-defined momentum but not have a well-defined mass or speed.

the nucleus in that cloud turns out to be just the right distance to counterbalance the force of confinement against the "force" of uncertainty. Any closer and the UP would cause the electron to flee away; any farther and the electrical force would pull it in closer.

(There is some evidence that human personality, the ego state, also undergoes multiple splits if confined. Adults with multiple personality syndrome have universally been found to have undergone severe childhood abuse, usually sexually oriented. When this occurred the fragile childhood persona divided into distinct personalities. Perhaps this is more than a metaphor, and multiple personality could be attributed to quantum behavior. We will examine this again in Part Six.)

In this manner the electron evades the nucleus even though classically speaking it would have to deal with it. The electrical force grabbing the electron tends to confine it to a region that would be singular—a region of infinite curvature and zero radial separation. But the cloud exhibits no singular behavior in spite of the force that is with it.

(Going back to the multiple personality, the child evades further abuse by splitting into multiple egos. In this way no one ego has to deal with the offending parent totally. The burden is shared among the troop.)

In a similar manner Hawking believes that there was no singular behavior at the time of the big bang. Even though an infinite number of bangs occurred, they all followed the laws of quantum physics and, without the presence of an observer, appeared in parallel universes. Equilibrium between the forces of the UP and the forces of attraction existing between all particles of matter (gravitationo-weak-electro-magneto-strong were all one force when that happened)* created

*There are four forces recognized in physical nature. They are called gravitational, electromagnetic, strong, and weak. The first two forces are well known. Gravity holds all large objects, such as planets and people, in its sway. Electromagnetic forces move electrically charged particles around, such as electrons in the copper wires hooked to your reading lamp from the nearest outlet. Weak and strong forces exist inside the nucleus of every

a quantum scenario similar to the first atom, the hydrogen atom—only instead of just one force being in contention, they all were.

Looking closely at the big bang, one sees that it had a structure—something like that of an atom, but yet different. Instead it appeared to have a structure similar to that of a black hole. Atoms can exist with unique energies called energy states. And one sees that this structure, following the laws of quantum physics, was capable of existing in unique quantum states. These are totally analogous to the quantum states that exist in the hydrogen atom. Accordingly there is a state of lowest energy—a ground state.

---

atom. Weak forces are responsible for a certain type of particle emission such as the decay of a neutron into a proton, an electron, and an antineutrino. Strong forces hold the nuclear particles inside the nucleus to each other. We believe that all of these forces were united into a unified field of force when time began.

Chapter
19

THE GROUND UPON WHICH THE UNIVERSE STOOD

A ground state in an atom turns out to be a configuration in which the inhabitants of the atom arrange themselves so that they have the smallest amount of energy possible. But to do this they must also be prepared to give up some objective quality, such as their locations with reference to each other. A similar situation occurs for a beginning universe. It seems that one alone can't be stable without a family of others surrounding it. Thus a universe is much like an atom, as I believe some old wise ones used to say when they looked at a grain of sand and saw in it the whole universe. Here we look at the analogy of an atom and a universe more closely to see how parallel universes began.

An Electron in an Atom and a Universe in a Universe of Universes

In the hydrogen atom, the two cohabitants are the electron and the proton. The pair cannot be thought of, however, as we normally think of two objects existing with each other. For example, the objects do not have defined locations within the atom. Instead both objects must follow the laws of quantum physics. Thus both exist as a single cloud of probability.

Even though both exist and comprise a single cloud of probability, because the mass of the proton is so much greater than the mass of the electron, one can deal with the electronic part of the cloud separately from the protonic part. The electronic part is spread out much farther in the space of the atom and, when atoms are grouped, actually constitutes the hardness, shape, size, and chemical properties of the atom. The proton's cloud is much more centralized and confined to the tiny region of the atom's center—its nucleus.

Thus when we think of an atom's cloud we are normally thinking of the electronic cloud. That cloud can appear in any one of a distinct set of patterns. These patterns change according to the energy of the electron in the atom. The lowest energy state's, or ground state's, pattern is spherical in shape. Higher energy states have different shapes, some spherical, and some with lobes sticking out in different directions. There are an infinite number of patterns according to the infinite number of possible energy states of the electron.

When an electron changes its energy—absorbs energy or emits it, the cloud instantly changes its shape and the atom instantly changes its chemical potency. Thus an energetically excited atom can react more efficiently than a ground state atom with another atom. Whenever an atom absorbs or liberates energy we say that the electron undergoes a *quantum jump*. Actually, it would be more accurate to say that the electronic cloud undergoes a quantum jump, for the electron cannot exist as a single point whenever the atom changes energy states.

It is also possible for the electron to exist as a single point in space for just an instant. However, according to the uncertainty principle (UP), it cannot then possess a unique energy. One says that it possesses a *superposition* of all possible energies at the same time.

Another way to say this is to say that the electron exists in parallel *energy* universes. In each of these universes it exists as a cloud with the appropriate shape and size consistent with the energy it has in that universe.

The Universe at Time Zero: Energy or Location?

In much the same manner as an electron behaves in an atom,* the universe itself, when it began, could have begun as a single point, in which case it would not have had a unique energy—or it could have begun in a unique energy state, in which case it could not have had a unique configuration.

If it began in its ground state, then it would not have had a unique radius or time of beginning—in other words, it would not have begun as a singularity, as a point. It would have had a structure, however, like a black hole (we have examined this structure in Part Three), but instead of a singularity at the center of the black hole, there would have been an opening, a topological space, connecting the universe to other possible energy universes.

Such openings are called *wormholes* by quantum topologists, physicist John Archibald Wheeler being the coiner of that term. If the universe is in its ground state, then it cannot spill out any energy. It cannot quantum leap to a new config-

*See "The Quantum Magic Lantern Theater" in Chapter 3.

uration unless energy from outside the universe enters. However, that energy entering from the outside would have to be exactly the energy needed to excite the universe to the first excited state. Any amount of energy flowing in would flow out if it did not have this exact value. Thus a ground state universe would be stable.

If it is in an excited state, however, it can, and indeed it will, emit energy, resulting in a catastrophic depletion of energy. This energy will vanish from our universe via the topological wormholes connecting it to its parallel universes. Presumably if it is in an excited state, it will drain energy to another parallel universe that is in its ground state.

Hawking points out rather elusively that it may not be only energy, but information, that leaks between the universes. In fact the ground state that Hawking refers to may be something in addition to a state of energy—it may be a ground state of information and order.

Hawking, in other words, is applying quantum rules where there were none to start with. We know that anything observable according to quantum rules can be arranged in a hierarchy from a ground or lowest value state to a highest state. Momentum, position, energy, angular momentum, spin, and mass are some of the observable qualities that quantum physics deals with. Hawking has added order or information to that list.

If our universe is in an excited state, not only is it possible for it to deexcite with rather catastrophic results, but we are in no position to predict when such an event would even take place. Information coming in from other universes would tell us when this would happen. But as Hawking cryptically informs us:

> *God may know what this information is, we don't....*
> *If the universe is not in a ground state science cannot predict the universe. The rest is up to God.**

*See Part 4, note 3.

Chapter 20

WHO SAW WHAT WHEN?

Thus Hawking brings us back to the question of the first observer. Without this observer, it is not clear how our universe began. In fact, it is increasingly clear that such a question cannot be decided. If our universe is really in a state of energy—ground or excited—then it must, following the atomic analogy, exist in an infinite number of parallel *position* universes.

Thus God, being the first observer, and saying, *Let there be light,* was probably misquoted. God more than likely said, *Let there be energy,* and with those words our universe, and all of the other infinite *position* universes, appeared simultaneously, filling *all-space* in *all-time.*

The Measure of All Things

Parallel universes contain information that must exist in order to produce all of the possibilities needed to create matter. These possibilities are measurable; some are more likely, rational, and meaningful than others. To remove all doubt consistent with uncertainty, these possibilities must have a numerical measure. Without the numerical measure of possibilities, and their ability to cohere, there would be no universe.

The uncertainty principle, with the connection it establishes between matter-energy and its location in time and space, shows that information has a rather special character—it is decisive and meaningful only if one considers probabilities with numerical values *between* zero and one.

Now this is not the case in a single universe of classical physics. There everything is or it isn't. Probabilities are either zero or one—nothing in between. Of course, we do assign probabilities between zero and one to situations we don't know about or can't practically predict. However, these probabilities are only measures of our ignorance or laziness.

Classical Gas on the Mind

For example, each particle in a gas—such as the air we breathe—classically speaking, occupies a unique spot in space and has a unique value of momentum. The UP, of course, forbids this. However, in many applications, it is useful to assume that real gases behave this way because the uncertainty introduces no real problem.

Assuming this, since there are vastly far too many particles to count even in the corner of a small room, we assign probabilities, calculable with the aid of a branch of physics called *statistical mechanics*, to deal with the classical properties of the gas. We really are ignorant of the exact location of each gas atom. However, each gas atom is imagined as if it does occupy a unique point while at the same time possessing a unique momentum.

We deal with our ignorance statistically. We say that any particular gas atom has a small probability of being right at the corner at any instant. Actually if we could watch it and classical physics was correct, we would see that it had a unique location and such a probability would be—for it being at the corner—either zero (it wasn't at the corner) or one (it was at the corner). In principle, things are where they are in the classical world and certainly nowhere else.

Classical Coin Flipping

Take another simpler example. If I flip a coin and catch it, it, being a classical object, will have either the heads or the tails side up. I may not know this, so I assign a probability of 0.50 to the prospect that the coin has heads up. However, the coin either has heads up or it doesn't. In other words, as far as the coin is concerned that probability was never ever 0.50—it was either 1.00 or nothing regardless of when, if ever, I looked.

Quantum Spin Flipping

But a quantum object—one that behaves like the spin state of an atom, for example—is quite different. If the probability of its having spin-*up* is 0.50, then the probability is 0.50 and not zero or one. It is not in a well-defined spin-*up* state. It doesn't really exist with spin-*up* if the probability for it to be spin-*up* is 0.50, for, as it turns out, it exists in a spin-*sideways* state, with certainty in this case; and according to the uncertainty principle it cannot have both spin-*up* and spin-*sideways* at the same time.

Thus in the quantum universe, significant things must exist with probabilities between zero and unity. But what could these things be? How are we to imagine them? In the usual or Copenhagen interpretation of quantum physics, we say they exist as *quantum wave functions* until they come into contact with some classical observing system. And then a miracle occurs, and they exist with probabilities equal to zero or one.

Of course, it is not clear what is meant by a classical observing system, since everything follows quantum rules—including the classical observing system.

But putting that to the side for the moment, we are still in a quandary because somehow, when an observation takes place—that is, there is an intelligence able to determine when something is or isn't present—then that something is or isn't present. If that intelligence cannot decide or is incapable of determining if that something is or isn't present, then that something must somehow be present with a probability that is between zero and one.

Now what can that mean? If it is only here, say, 50 percent of the time, where is it when it isn't here? The answer is, according to this interpretation of quantum physics, in a *parallel universe.* And there it is also present with only 50 percent of its beingness.

What happens when such an intelligence—as is able to decide that the thing is here or is not here—observes the quantum object? Then does 50 percent suddenly change into 100 percent or nothing? If so, we are right back to square one again—a miracle occurs. The way out is to say that when that intelligence appears and observes, that intelligence also splits, part of it appearing in one universe and the other part in the parallel universe. Thus in both universes an answer appears. In one the thing is seen and in the other it is not seen.

Thus in the example of the spin-*sideways* particle, when an observer puts a spin-*up* measuring device in the lab and carries out a spin-*up* measurement, the observer splits into two alter egos as the measurement is being carried out. One of the observers sees spin-*up* with certainty and the other with zero probability, and each feels that he or she is the only observer.

Thus in the beginning when the universes all began, God was split and His/Her intelligence was parceled out into myriad parallel *position* universes so that a single *energy* universe, an energy state universe, could manifest itself. If so, then God got caught by Her/His own creation, and much like Br'er Rabbit, got entangled with the Tar Baby of Uncle Remus's tale. And the universe was created. And there was intelligence in it. In Part Five we will see how this intelligence changed our view of time.

PART FIVE

HOW PARALLEL UNIVERSES PREDICT A NEW NOTION OF TIME

If you ask me what time is, I don't know the answer. If you don't ask me, I do.

Saint Augustine

This is often the way it is in physics—our mistake is not that we take our theories too seriously, but that we do not take them seriously enough.

Steven Weinberg,
Nobel Prize-Winning Physicist

What did the early universe, the universe that existed before there were any observers, look like? Since there were no observers at the time of the big bang and since quantum rules run the show, there must, and I emphasize the word *must,* have been parallel universes occurring then, because all possible scenarios for the big bang must have occurred according to the rules of quantum physics.

Even the conservative Copenhagenists should agree. The world we see, according to the Copenhagen physicists—who believe that an act of observation rules out alternatives—appears when an observation takes place. Even they would agree that before the first observation, all we can really say is that the universe was in some superposition of quantum possibilities.

Take, for example, the radius of the early universe. Did it even have one? How could it, because according to a quantum picture it wouldn't have had any radius until that radius was measured? Who measured it? When was that measurement accomplished? Several physicists have addressed these questions and have come to a startling conclusion—one that is nevertheless consistent with the theme of this book: It is our observations now that are determining the past.[1]

Do Thoughts and Wishes Time Travel?

Thus an observation of an event now somehow sends a message backward in time and "causes" events in the past. If this is true, then what is really the past? It would seem that there is no absolute past, because there is always the possibility at any time that some present event will alter it.

A way out of this paradox is found in the parallel universe theory. Accordingly, there is no fixed past. The past we believe is *the* past is what a community of communicating intelligent beings choose to be the past. Other pasts are *out there* waiting to be discovered. In other words, there are parallel pasts—an infinite number of them. The past that is altered by the present is just one of the many.

Since, according to relativity theory, there is no such thing as the absolute present, then what is present for someone could be the past or the future for another. Consequently, it would seem that the future also communicates with the present. But which future? The future that we believe will be the future is again that which is chosen by a community of intelligent communicating persons. According to the quantum rules governing parallel worlds there are an indefinite number of futures. So how can *a* future that is not fixed communicate with the present? Which future sends messages back to us?

The only consistent view possible is that all possible futures act on the present. Viewing the whole scenario of an infinite number of parallel universes as one big continuum stretching from the infinite pasts (actually not so infinite—only about 15 billion years ago) to the infinite futures, the effects of observations propagate in both directions through time—to the pasts and to the futures. What is future or what is past is purely a personal viewpoint, like being on one of many roads in a gigantic city leading from somewhere to somewhere.

And here comes a surprise. If the future communicates with the present, and by the same line of reasoning, the present communicates with the past, then it must follow that time is not fixed. We are not stuck in it like so many flies in a jar full of jelly. If we are not stuck in time, can we, like the unsuspecting hero of Kurt Vonnegut's *Slaughterhouse Five,*[2] become unglued in time? Is a time machine, a device that can send one backward or forward in time, possible?

With parallel universes appearing as a result of spacetime distortion induced by gravity, new effects appear. For example, assuming that travel to a parallel universe is possible, it is impossible to go to it without traveling in time. In other words, time travel as described by the multitudes of science-fiction writers now turns out to be possible. Just get near a whopping spacetime warp and you have found one of nature's time machines.

Now the existence of parallel universes is troublesome enough. Why bring in time machines? Again the answer is consistency. As strange as it appears at first glance, time machines—devices that enable a conscious being to travel in time backward or forward—must be constructable if parallel worlds exist.

Do time machines already exist? Assuming that the general theory of relativity is correct, then time machines and parallel universes must exist. If they don't, Einstein was wrong. And if Einstein was right, they exist inside those stellar-collapsed objects called black holes. It is here, inside a black hole, where parallel universes and relativity really begin to fit together. Massive objects confined to small regions of space produce curves in spacetime. Curving spacetime introduces gravitational distortion, and such distortion creates parallel universes and time machines.

Nearly any subnuclear bit of matter may be capable of such shenanigans. In particular, the spinning bits of matter called electrons may be dense enough to be black holes. If so, they lead to an infinite number of parallel universes, since all black holes lead to parallel universes.*

Certainly electrons are possible candidates for message receivers from the future. They are small enough, being as far as any one can tell, point-sized particles, and possessing spin, so that they may qualify as holes leading to other parallel worlds. This may be why they are capable of existing as clouds in atoms.

If we could enter an electron, we might see just the kinds of things that might be seen when entering a spinning black hole as predicted to exist by general relativity. So far this is just a speculation, of course. If true, then it would be very possible to witness other worlds by somehow tuning in to our own electrons. Maybe a day will come when this will be possible by some highly developed technology—an electronic form of biofeedback. Then our bodies would themselves be time machines.

Time machines can be constructed consistent with the laws of quantum physics and general relativity, provided parallel universes indeed exist. In other words, if there are parallel universes there must already be time machines present: those made from the burnt-out astrophysical thermonuclear ovens of nature—neutron stars and black holes.

However, there are no time machines yet built by humans. Nevertheless, the existence of parallel universes—those that appear as a result of the observance of any physical system,

*But you wouldn't want to enter a stagnant (nonrotating) black hole. It rips everything apart. A spinning black hole doesn't do that, if one enters it carefully.

according to the quantum rules—makes it possible to communicate with both the past and the future through the actions of the quantum wave of probability. It turns out this wave can travel both forward and backward through time.

But wouldn't communication with the past and/or future involve many paradoxes or logical inconsistencies? A typical inconsistency is: if time travel were possible, then I could return to my grandfather when he was a tot and accidentally cause his death. Or I could perform some experiment that would propagate back to grandpa's day and convince him not to marry grandma. How could I then ever have been born in my own time period if granddad never made it with grandma?

The way out of the paradox is simple enough. In that parallel universe where granddad never made it to the first grade or ran off with the circus, you were not born. But there are other granddad universes and therefore other ways to make you.

Assuming that it is possible to escape the time travel paradoxes that result when a person goes back in time and alters the past, the possibility exisits that the future and the past are already communicating with the present. The tiny quantum, through the reality of the quantum wave of probability and its behavior in spacetime, already implies that information can flow from past to present and from future to present. Thus it implies the existence of both the past and the future "simultaneously" with our own time.

Neutron Star Time Machines

Can we actually build time machines? General relativity, through its equations, contains the means by which a real physical time machine could be built. However, we would need an enormous amount of energy to do this. Basically, make a long enough cylinder out of already existing rapidly rotating neutron stars and you will have a time machine. Such a cylinder contains in its immediate neighborhood concentric zones where time goes backward.

In the final chapter of Part Five I'll take you on an imaginary trip to the future where such a time machine is built and time travelers are about to start out on the first physical time travel. But first we need to look at time travel and its paradoxes.

Chapter 21

TIME TRAVEL

Time travel has always captured our imagination. Who among us doesn't wish to find himself or herself boarding an American Airlines or TWA spacetime vehicle that would take us on a journey either back to the past or forward to the future. The prospect of being tourists during the time of the crucifixion of Christ or the teenage years of the Buddha, or even closer to the time of our early childhood, appears quite compelling. Maybe a journey to the future where diseases, such as cancer, are wiped out or where the human life-span is five hundred years or more is our desire.

In a recent movie blockbuster, *Back to the Future*, a teenager is given the opportunity to return to the time of his own parents' teenage years, a time before his parents had married, and, of course, he wasn't born yet. Finding that his father-to-be was a nerd and that his mother-to-be hadn't the slightest interest in dad, our hero is forced to do something about it or else he doesn't get born. Complicating the plot is the fact that his mother-to-be finds her unborn yet quite present and handsome teenage son quite attractive, and...I won't reveal the whole plot.

Nevertheless, the story is, as typically these stories are, filled with amusing paradoxes. Suppose his parents never actually marry. Will he exist? What will happen if he then returns to his own time aboard the time machine? Will he return to a world in which he never was? Then what?

Even Pepsi-Cola got into the time machine craze. One of their many television advertisements had a young man entering a time machine to return to the past. Only he inadvertently brings along an aluminum can of Pepsi with him. The scientists in charge watch him disappear in the machine as the youth transports back to the early 1930s. When they notice that he took a can of Pepsi with him, they worry about its presence in that time period. As they walk out of the science laboratory, they say to each other, "Don't worry about the can. What harm could one can of Pepsi do?" As they continue on their journey, we see the Coke machine in the corner of the lab dissolve into nothingness, and following the two scientists out of the building we see the local Coca-Cola factory dissolving away—the message being that once today's Pepsi is discovered in the early thirties, the new Pepsi-Cola is by far the choice of that generation, so that Coca-Cola never really gets off the ground. Then comes the logo, "Pepsi, the choice of a new generation."

The implication is that if a factor in the past is altered by knowledge from the future materializing in that past, it will have great ramifications in the present, perhaps wiping out whole present realities.

One can see why this could happen. If you just think of the fact that you are the product of two parents, who are the products of four parents, who are the products of eight parents, who are the products of sixteen parents, and so on, doubling with each backward-through-time generation, we see that any alteration of a distant foreparent could lead to your not being born at all.

Suppose you were to time travel to the future. Here there doesn't seem to be any real problem. Nothing in the future could affect your present. So it would seem. But suppose you come back from the future and alter the present with the knowledge you have from that future. Then it appears that you could change the future you just returned from.

All of these paradoxes arise from backward-through-time information travel. You could go back in time, enter the mind of your grandfather and tell him not to make love that night with your grandmother. Consequently your mother never gets born, and neither do you. Or do you? After all, you made the time journey in the first place, so you must have been here to have gone back in time to there.

Assuming that information could travel backward in time, how can such paradoxes be resolved? The answer is, there must be parallel universes.

Chapter
22

TIME TRAVEL PARADOX RESOLUTION BY PARALLEL UNIVERSES

The ordinary paradoxes of quantum physics are many. How can a single subatomic object, such as an electron in a hydrogen atom, occupy more than one place at one time? The answer is, of course, get rid of time. Now there is no paradox. If we ask how can an ordinary object occupy more than one place in a given sequence, we have no problem. It can, for example, do so by occupying a different place at a different time. I have occupied the penthouse suite at the Hilton Hotel in Calcutta, my chair at the office in the university, my chair at the office of the firm where I consult, and the streets of Tokyo where I jog, all in the same sequence. But not at the same time.

Quantum physics rids us of time because it can do nothing with it other than use it as a sequence-of-states ordering parameter. But since it is a useful concept, in the sense that we can create futures which are less-suffering than the past, such as traffic lights that order the future flows of living beings through the same space, we use it. And we then try to order up the rest of the universe following our imagination-created concept.

Using time as God-given, we have created the laws of physics. This was done quite explicitly by God-fearing Sir Isaac Newton in the black plague days of England's seventeenth century. Our own twentieth century "tuned" in to Newton's mind (albeit well aided by Newton's writings) and extended those laws to our present quantum condition where we are just able to see around the corner to the next parallel universe of the twenty-first century.

Consider the grandfather paradox once again. Suppose you travel back in time and accidentally cause the death of your grandfather before he reaches puberty. You can't get born because your mother or your father can't get born. How is this resolved by parallel universes?

The answer is in that parallel world sequence where you and your prepubescent grandfather "resonate" quantum wave streams ending his young life; you are not born. Thus in that world you don't exist. In this world, which parallels that one, nonexistence occurs only as a fanciful thought. In that world, some other "you" perhaps thinks of you as not existing.

Thus all annihilation backward-through-time paradoxes are resolved by the existence of parallel worlds, universes where the originator in the future does not exist. It is much like a transfer from one train to another at a train station. You board the train in the present, go back in time and alter an event, having direct and causal bearing on the existence of yourself, and return to the future. The future where you are not is on another track.

The only criterion for existence in all of this is self-consistency. Think of these time travels as time-loops.* The loop

*Time-loops are journeys that loop from present to future and then backward to the present, or journeys that start in the present and go back in time and then return to the present, or any combination thereof.

where you and your child-grandfather (who wishes to grow up) resonate the strongest is the existence associated with your child-granddad picturing you in the future and you remembering him. In the next chapter we will see how this works from the quantum physical parallel universes perspective.

Chapter 23

CLASHING WAVES OF TIME

As we proceed through our everyday lives, day after day, moment following moment, we seem to be on a kind of treadmill. The ground under our feet appears to be rushing forever backward as we go on, hanging in there, keeping up the pace, and seemingly never getting anywhere. After all, each one of us is older today than yesterday, whether any one of us likes it or not.

Yet there is something timeless in this aging. I "feel" the same at this moment as I did moments ago. Since I can say this at any time, including those moments that were ago, I conclude that I am the same and that nothing associated with "me" has really changed. I am told through my senses and my observations of the world around me that time has marched on. But inside of "me" there is no time that I can directly perceive or even measure in terms of something actually making a difference.

I know that I could go to my local physician and ask him for a physical examination. Undoubtedly, he would find that my internal organs have changed, and like every aging human being, signs of aging would appear. Yet, I contend, I "feel" no such signs. Inside of "me" exists a timeless condition—something I associate as the seat of my consciousness, my ego, or my timeless "I."

From the point of view of quantum physics, time, too, is timeless, for quantum physics deals with observables—things and their states which are perceivable. For example, a ball in a baseball park is observable. The ball's states are its possible locations inside the ball park. Even if the ball is struck by this year's home-run champ, and the ball leaves Shea or Dodger Stadium and lands over the wall in a neighbor's backyard, the ball still has a state because it still has a position.

There are many such observables in quantum physics. Electric charge, momentum, spin, angular momentum, mass, energy, and more. Each of these observables possesses states—possible perceivable experiences that can be quantified. But conspicuously missing among them, even though it would appear "obvious" that it should be there, is time and its "states."

Time Is Invisible

There is no observable corresponding to time in quantum physics. Thus from the point of view of the new physics, time plays a different role from the other "observables" because it cannot ever be directly observed. One might wonder just what one is observing with a casual glance at a wristwatch. Although a clock indeed "tells the time," it is really the motion of an object—the rotation of the earth, or the movement of our planet around the sun—that is being observed, for time has no direct measure. One must infer the time from the movement of a second hand "sweep," for example.

A length of an object has a direct measure. It can be repeated and verified. We can't do this with time. Once passed, that's it. There is no going back and repeating the measurement. A real measure of time would mean that one could travel back and repeat the measurement as many times as one liked.

If one looks at the fundamental or most basic equations of classical or quantum physics,* one is immediately struck with this fact. Nowhere is there a value of time associated with an observable called time. As far as the equations are concerned, time is just a convenient ordering parameter—a way of keeping track of things placed alongside each other in a sequence. We mark all observables according to where in "time" they are likely to occur in quantum physics equations, or where they will occur in classical physics equations.

Furthermore, it makes absolutely no difference whether we regard this ordering parameter as proceeding from zero to infinity (i.e., from low numbers to high numbers) or from infinity to zero (from high numbers to low ones). The equations of physics do not reflect any particular time order.

Classical Physics Has No Time Order

If the world we observe were to actually follow the equations of classical physics, there would be processes which we could observe that would contradict our usual "sense" of time order. We would see objects that normally rise, fall. Objects that normally go toward the right, proceed toward the left. For example, a ball lying on the floor could suddenly begin hopping, with each hop getting higher and higher until it just leaped into my hand. While this would certainly look strange, there is nothing in the classical laws of physics to prevent this from happening. However, looking at the most

*These include Newton's laws of classical mechanics and Schrödinger's equation—the master equation of quantum mechanics.

microscopic phenomena, observables that we are usually not accustomed to watching, we probably would not be struck by anything strange going on. I mean, who cares if that ameba spits out that piece of food or not.

But watching a hungry man eating chicken, proceeding backward-through-time, would indeed be a strange experience. We would see him pick up an empty bone from his plate, shove it into his mouth and then watch him chew. As we glimpsed inside his mouth from time to time, as we usually are prone to do when we watch a hungry man eat, we would find that pieces of meat would be adhering to the bone, layer upon layer, rather than being pulled from the bone by a voracious appetite.

As soon as the man was "full" he would stop eating, pull the bone completely covered by hot chicken meat from his colder mouth, and put the hotter assembled chicken leg down on the plate. Of course, the man would appear a little strange after his meal. He would actually look hungrier after "eating" than before he started. And most likely he would be hungrier, although I don't really know that, for the man eating backward-in-time would be simply time-ordering his meal in the opposite sense from our time order.

Imagine a person living backward in time day before day. If he walked through the streets of San Francisco, he would appear quite strange because he would be walking backward. Only if he decided to climb Nob Hill, step before step, looking in the direction from which he came, would he appear to the outside world of sophisticated San Franciscans as normal. They would see him walking down the hill, even though he would actually be climbing it with his back toward the incline.

I am sure the reader can imagine even more bizarre scenarios for our time-reversed hero. Such scenarios could actually occur if the world were completely described by the classical equations of motion. Of course, most sequences involving large numbers of events, such as a ball spontaneously hopping off the floor, would appear highly unlikely even in classical physics. Broken eggs do not reassemble and jump off the floor into a tipping bowl on the table. Such sequences usually involve a degree of randomness associated with the lack of specification of the initial or boundary conditions associated with them. If we could specify the egg's broken shells, bits of white and yellow matter, and bounce the floor with the right amount of energy and momentum, nothing in classical physics would prevent the egg from reassembling and doing its time-reversed thing.

The air molecules in a room, for another example, could all speed to the nearest corner, leaving the rest of the room devoid of air, if those initial conditions were designed in the right way. However, given a random distribution of the speeds of the molecules, we are assured that none of us will suffocate. My point is that classical physics appears to contain the experience of time order we all have come to know as the "march of time," not because it does contain a time order, but because the boundary or initial conditions governing the behavior are largely random and uncontrollable. Even though the classical equations would allow time reversal to occur, it is statistically not the case that those conditions that would send all the molecules in the room to the corner would ever occur.

Quantum Physics Has No Time Order

Whan quantum physics is used, a similar situation arises. Again time order is arbitrary, and sequences of "events" can arise in which a reversal of time order occurs. The events in quantum physics equations are, however, not existential as they are in classical physics. In classical physics an event is described by one possibility and one possibility only. The event actually occurred, will occur, or is occurring. In quantum physics the "things" calculated have not occurred, they will not occur, nor are they occurring. They have not taken place in the past, the future, or the present. They are ghosts

—probabilities of what was, what is, or what will be.

These "things" are quantum wave functions. They describe distributions in space of possible occurrences of real events. To do so these quantum waves not only must be nonexistential, but must also be capable of traveling both forward and backward through time and be able to link one parallel universe with another.

Time Waves

In two recently published papers,[3] physicist John G. Cramer, from the University of Washington in Seattle, brings into play a new and exciting idea. Cramer reminds us that before any measurement takes place upon a quantum physical system, the system is represented by a mathematical expression —the quantum wave function. The quantum wave function represents many possibilities simultaneously.

We can imagine this function as a wave traveling through space. All points on the surface of the wave represent places where an event is likely to occur. When any observation occurs, the wave is imagined to collapse, like a popped balloon, from a surface of possible events to a single undeniable fact.

The big question, which is also the measurement problem referred to earlier, is: How does this collapse occur? A clue lies in the manner in which the physicist computes numbers that measure the probabilities from this wave. In order to compute the probability of this collapse taking place—in other words, the probability of the event associated with the collapse—the wave must be *multiplied* by another wave that is similar in form and content to the original wave. This wave, for mathematical reasons, is called the *complex-conjugate* wave.

Multiplying two mathematical entities together to obtain a single number is quite common in physics. For example, in classical mechanics, the force on an object is found by multiplying its mass by its acceleration. This rule of multiplication follows from the second law of motion given by Isaac Newton.

However, even though quantum physics is quite rigorous, nowhere in it is there any law explaining what is physically occurring when one multiplies a quantum wave by its complex-conjugate. Nowhere is the complex-conjugate wave given any physical significance. Except for a funny little quirk. The complex-conjugate wave is also a solution to the equations of quantum physics provided that in writing those equations you let time run backward instead of forward.*

Now the quantum wave itself, although physicists have a great deal of faith in its existence, has never been seen. It is just a solution to an equation. But if the quantum wave is a real physical wave—one that exists and propagates through space and in time—then the conjugate wave, which also has never been seen, is not a mystery, provided you are willing to borrow an idea from science fiction and let it run backward through time. So goes the argument. If the quantum wave is a real wave, then the conjugate wave is also a real physical wave,† but with a twist of time.

Now any wave, including the quantum wave, moves from one place to another. So it must take time to do that. We can imagine the wave propagating through space much as a ripple from a stone dropped in a still pond. We picture it expanding ever outward.

However a time-reversed wave would not do that. It would just suddenly appear at the pond's boundary, and, as if we were watching a movie of it run backward through the projector, it would squeeze in on itself, ever contracting until it all collapsed to a single place where the stone hit the still water.

*This is not the first time that someone noticed that running the clock backward in physics equations could lead to a new discovery. Richard Feynman received the Nobel prize for his use of this idea in the study of the interactions of photons and electrons called *quantum electrodynamics*.

†I have called the complex-conjugate wave the *star-wave* in my previous book of the same title.

Thus the conjugate wave travels in the opposite direction as it goes back through time. And as it does so, it meets up with the original wave. In the physics of waves, the conjugate wave is said to *modulate* the original wave.

Now the idea of wave modulation is quite familiar to scientists and engineers working in the area of radar, radio, and television. When you tune your receiver, television set, or radio to a station, you are picking out of the air a certain well-defined and quite narrow band of transmission frequencies sent by the broadcasting stations. The central part of this band is called the *carrier* frequency. However, that carrier frequency is not what you hear or see. The sounds you hear and the pictures you watch are carried piggyback by the carrier wave. The information you see and hear is simply wave forms that modulate, or cause the strength or the frequency of the carrier wave to change.

Similarly, the conjugate wave modulates the original wave, and mathematically this is nothing more than the multiplication of the two waves together.

Thus, Cramer explains why probabilities are computed the way they are, by the multiplication of the original wave by its complex-conjugate wave. In order that any event occur, both quantum waves must be simultaneously present, one modulating the other. Here, then, is Cramer's explanation of the wave function collapse—it occurs when the future-generated conjugate wave propagates back through time to the origin of the quantum wave itself. There the two waves multiply, and the result is the creation of the probability for the event occurring at the site of the original wave.

Cramer calls the original wave an *offer* wave, and the conjugate wave an *echo* wave. He also calls the multiplication of the two waves a *transaction.* Thus a transaction occurs—involving an *offer* and an *echo*—much like the transaction between a computer and one of its peripheral devices, say, a printer, or another computer over a telephone line.

In these computer technology examples, an offer wave is sent to a receiver. The receiver accepts the offer and confirms it by sending back along the same line to the offerer an echo of the offer, indicating to the offerer that the message was received.

In the quantum wave–complex-conjugate wave sequence, the exchange is the same, except that the offer and the echo cyclically repeat until the net exchange of energy, and other physical quantities that will manifest, satisfy certain requirements. These include the conservation laws of physics and any other restrictions imposed on the quantum wave, known as boundary conditions. When this is all taken into account, the transaction is complete, and all's well that ends well.

A New Picture of Time

If one takes Cramer's interpretation seriously, then we have a whole new picture of time in regard to quantum events. Every observation is both the start of a wave propagating toward the future in search of a receiver-event and is itself the receiver of a wave which had propagated toward it from some past observation-event. In other words, every observation—every act of conscious awareness—sends out both a wave toward the future and a wave toward the past. Both the beginning of the wave and the end appear to begin in our mind—our mind in the future and our mind in the present. Two events in normal or serial time are then said to be significantly connected, that is, meaningfully associated, one with respect to the other, provided that the transaction between them conserves the necessary physical constants and satisfies the necessary boundary conditions.

Cramer emphasizes that the transactional picture is just an interpretation of quantum physics, and as such he doesn't expect that any new experimental evidence supporting it over any other interpretation will be forthcoming. He sees it as a way of understanding and developing intuition in regard to teaching quantum physics to students. It also helps explain quantum physics paradoxes that are quite difficult to explain if one insists that time runs only one way from past to future. We will look at one of those paradoxes and how the transactional picture explains it in the next chapter.

Which Future Sends the Message?

But is that all there is to Cramer's idea? There is still an interesting problem afoot. Which future sends back the echo wave? Cramer believes that only one future does this—the one producing the echo that happens to form a successful transaction with the present. However, we are still faced with the problem of explaining events that occur with probabilities less than the best chance. Here I would like to present a new idea. It seems to me that Cramer's ideas must be interpreted in light of the parallel worlds idea. All futures return the message, not just the best chance future. There are more futures "listening" to the broadcast than just the one with the most sensitive and powerful receiver.

If both the quantum wave and the complex-conjugate wave are real, then time must be different from a one-way river. Events that have passed must still be around. Events that will be must exist like boulders behind blind corners of the roads of life. And if both the future and the past exist, now, then quantum physics implies that devices must exist that enable us to tune in on the future and resonate with the past.

These devices may be our own brains. When we remember a past event, we are doing no such thing as digging through a file or memory bank like a computer memory. We are, following quantum rules, constructing or creating a past based upon the multiplication of two clashing time-order streams of quantum waves. Taking this literally, this means that the past stream, the one going from the past toward the present moment, must be originating in the past in the same way that the present stream, the one going from the present back in time toward the past, is originating in the present. This means that the past and the present must somehow exist "side-by-side."

Me in the Future Talks to Me in the Now

As strange as this thought may appear, it follows that the future, too, "exists" side-by-side with the present and that we at this moment are sending quantum waves in that direction of time (which means we are laying out possible sequences of existential scenarios in a present-to-future order). Someone called "me" in the future is also sending back-through-time quantum waves which will clash with the waves being generated here and now.

If these waves match, in the sense that the modulation produces a combined wave of some strength and there is a resonance, which means that the future event and the present event are meaningful for me, then a real future is created from our present point of view and a real memory of sequences is created in the future. If the two streams do not match, meaning that the modulation produces a combined wave of weak strength and there is no resonance, then the connection of that future and the present will be devoid of meaning. Meaning in this sense refers to probability. The greater the probability, the more meaningful the transaction and the greater the chance of it occurring.

The closer in "time" the sources of these waves are, the more likely that the two countertime quantum wave streams will "marry" and produce a strong probability—one that has a good chance of becoming real. It is quite possible that visionaries are those who successfully marry streams coming from far time-distant sources. People unable to cope with life are those who lack the ability to do this for the shortest time-distances with any degree of success.

The past, present, and future exist side-by-side. If we were totally able to "marry" corresponding times each and every moment of our time-bound existences, there would indeed be no sense of time and we would all realize the timeless state, which is taken to be our true or base state of reality by many spiritual practices. But we fail to do this because we fail to discriminate between the many past and future sending stations attempting to communicate with us, so we live time-bound lives disconnected, to some extent, from the past and the future.

We might wonder what we can do to pick up a better signal from the future. Well, if there are parallel futures out there, all broadcasting back in time to us, surely there are some people who *hear* them or *see* them. Perhaps people who have lucid dreams are able to sense them when nothing else is disturbing their major senses. Perhaps certain mental disorders are visions of the future. Even, perhaps, flying saucer sightings and "on-board" visitations of people with beings from another reality are more than hoaxes or the ramblings of simply misguided or disturbed people. I believe that visionaries are those who are able to turn away from everyday life concerns and tune in to these other worlds, whether they are past-life recalls from parallel worlds gone-by or future-life recalls from worlds yet-to-be.

Gone-by and yet-to-be are simply reference points based on our sense of now. They are simultaneous with us in the parallel worlds view of time.

These pasts and futures are, as I said, side-by-side, parallel universes. The past and the future which we remember and appraise as real are just those time wave clashes with greatest strengths and most resonance. *Now* can be defined as that event or sequence of near events that are meaningfully connected wave clashes. What we call "now" moments are those clashes of waves that are "in tune" with each other and have the greatest strengths.

In the next chapter we will see how time wave clashes can be used to explain a paradox: how the choice made now by an observer alters the past.

Chapter 24

WHEELER'S CHOICE

When the universe first began, no attempt was made to differentiate one possible evolving universe from another. All possible universes existed in a superimposed state acting as a single universe, according to the parallel worlds concept. Nothing was defined. The universe at this time did not possess a well-defined radius, for example. But then a mysterious interaction occurred. The universe was split into several, perhaps an infinite number of parallel outcomes, each specifying a radius for its universe. How did this happen? If we ascribe an observer in all of this, then we raise the obvious questions: who is the observer, and when did the observation occur? The answers may be, as surprising as it sounds, we and now.

By peering back in time, by looking out into the universe at light signals that were emitted millions and millions, perhaps billions and billions of years ago, we may be the observers that are causing the early universe to split, and thereby we are choosing by our observations today what the radius and other physical parameters of the early universe were.

This is an example of what visionary physicist John A. Wheeler calls "delayed-choice" measurements.[4] Accordingly, it is our choices made now in the present that determine what the past had to have been. Wheeler's ideas are quite profound and paradoxical, so I am going to use two examples to explain them. The first example is illustrative and clearly describes the paradox. In the second example, I'll use Cramer's transactional interpretation and the parallel worlds concept to show how the paradox is resolved.

A Photon from the Dawn of Time

Consider a single photon (a particle of light) emitted at the dawn of time during the big bang that travels a vast distance of about fifteen billion light-years from the edge of our known universe to reach our eyes now. This photon was emitted fifteen billion years ago when the universe was imagined to begin.

According to quantum physics, this photon could have traveled to our measuring instrument, either an electric or

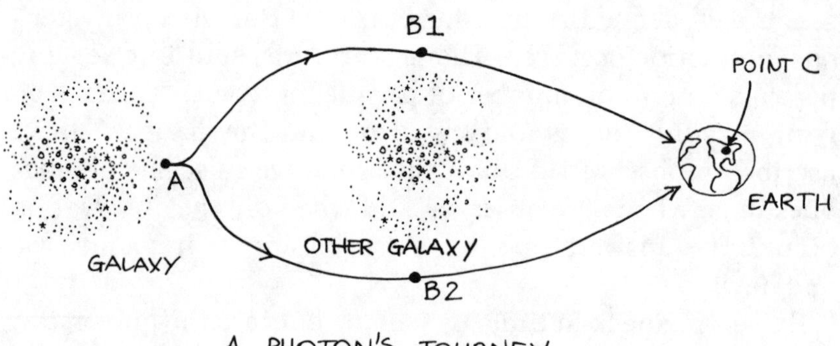

A PHOTON'S JOURNEY

human eye, via several alternate means. It could have, for example, followed a single path, going from point A to an intermediate point, B1, between galaxies and then to our instrument located at point C, back here on terra firma. Or it could have traveled from point A to another intermediate intergalactic point, B2, and then on to our instrument at point C. Suppose, for the sake of argument, that points B1 and B2 are themselves ten billion light-years apart. By setting our instrument so that it can catch the photon regardless of which path it followed, we can determine by which path the photon arrived. Call this experimental arrangement *setup one*.

On the other hand, if we set up the instrument in a different "mode," called *setup two*, so that we can no longer tell by which path the photon arrived, then the two paths would have flowed together like joining wave streams and superimposed on each other at point C. Quantum mechanically the two alternate paths, like different waves flowing together, interfere with each other so that the outcome at point C will be different in *setup two* from that in the first setup.

Thus it is we who decide, by our choice of setup now, whether the photon shall have traveled by *either path* or by *both paths*. And we make this decision at the very last minute of the photon's existence, even though that photon left its source fifteen billion years before we walked the planet.

Experimental Verification of Backward-Through-Time Travel

Although this sounds like science fiction, such a delayed-choice experiment[5] (using terrestrial photons) was actually carried out in College Park, Maryland, in 1985 by three physicists—Carroll Alley, Oleg Jakubowicz, and William Wickes of the University of Maryland. The Alley, Jakubowicz, and Wickes experiment was literally all done with mirrors. By electronically inserting mirror surfaces in strategic locations in their setup they were able to manipulate a single photon emitted from a source into their apparatus so that it would pass into two well-separated channels. After passing through the channels the photon would be observed in two different ways. By inserting the mirror just before the observation point, the photon could be made to interfere with itself. By leaving the mirror out, the photon would not undergo any such interference.

With the mirror present, their data confirmed that the photon had to pass through both channels simultaneously. Strange indeed, for there was just one photon present. With the mirror out, the photon would pass through one channel or the other. Even more bizarre, the decision to insert or not to insert the mirror wasn't made until after the photon had entered the setup. They indeed showed that their last nanosecond decision did choose by which means a photon traveled, either by both paths or by one of the two paths. This experiment confirmed Wheeler's "choice." And it affirmed that a last-minute choice can affect what we mean by the past.

Now Makes the Past

It seems extremely difficult to conceive of an undefined past that only becomes defined by our actions now. Yet quantum physics compels us to take this view. If all our observations of the early universe conform to *setup twos,* so that no attempt is made to differentiate one universal outcome from another, the universe remains "out there" and undifferentiated. It has no radius because it has all possible radii. In a sense, it has no beginning because no setup was made to "create" that beginning. Its radius is still caught up in the wave of possibility. But by determining through choices made today, using *setup ones,* we can "create" which radius the universe had (or should I say has had—my use of tenses becomes confused when the present can affect the past). Our choice now branches us onto a parallel universe where the radius of the universe at the beginning of time is determined.

Does the Future Influence the Present?

Let's consider the paradox again using the idea of clashing time waves. Consider a slightly modified version of Alley, Jakubowicz, and Wickes's setup—a simple earth-bound experiment consisting of a light source, a screen containing two slits, similar to the so-called double-slit experiment,[6] a light-sensitive film-emulsion screen mounted in back of the slits, and a pair of telescopes mounted in back of the screen. The screen is held on pivots so that it can be rotated into an up position to capture a photon, or rotated down so that the photon continues on its journey to the telescopes.

The distance between the emulsion screen and the slits is long enough to allow the experimenter plenty of time after the photon has passed through the double slits to decide whether to rotate the screen into position or to leave it down. By long enough, I mean that the operation of the rotation of the screen into position can occur after the photon passes through the double slits and before it reaches the telescopes. It is important that the experimenter's choice whether to rotate the screen into position or leave it down takes place after the photon has passed through the double slits.

If the screen is left down, the photon continues on one of its paths and finally ends up reaching one of the two telescopes. Each telescope is mounted directly along a line-of-sight in back of each slit. Thus if a photon hits telescope 1, the photon must have traveled through slit 1, and similarly for telescope 2.

The experimenter, in making the decision at the last moment whether to rotate the screen into position or not, finds himself in Wheeler's quandary. Suppose that he places the screen into position. Then, according to quantum mechanics, the single photon must pass through both slits simulta-

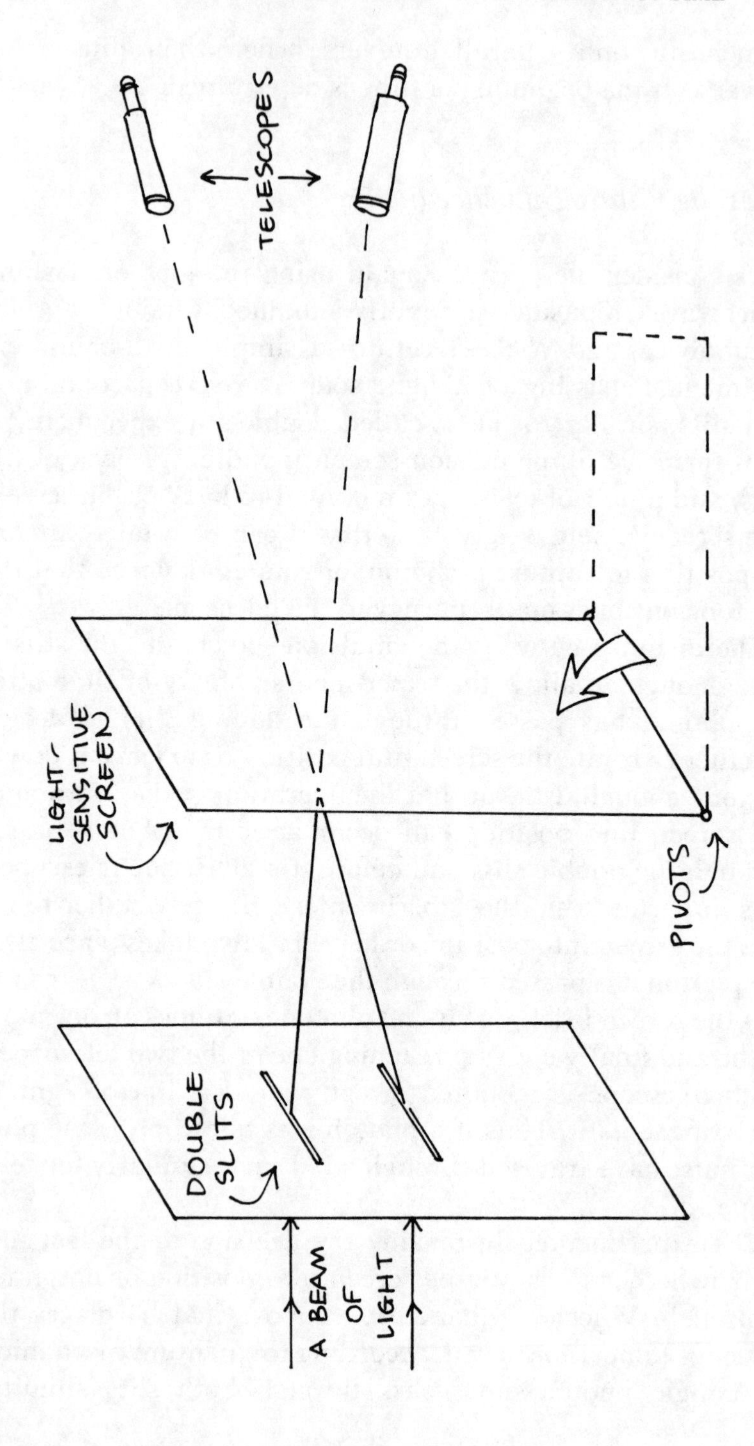
TELESCOPES
LIGHT-SENSITIVE SCREEN
PIVOTS
DOUBLE SLITS
A BEAM OF LIGHT

neously in order to make an interference record on the screen film. In other words, the film-emulsion when rotated into the up position acts as a device for determining the wavelike property of the photon.

On the other hand, if he decides to leave the screen down, the photon reaches one telescope or the other, indicating that it passed through one slit or the other. Thus with the screen down, the telescopes measure the particlelike property of the photon.

Screen up—the photon passes as a wave through both slits. Screen down—the photon passes through either slit as a particle.

But since it is well after the photon has passed through the slits that the screen is rotated into position or not, the experimenter chooses which is the legitimate past for the photon—one path or both paths. In other words, the experimenter delays his choice until the very last instant, and then that choice determines whether the photon traveled as a wave or as a particle.

The effect, in other words, has come before the cause. The cause (the experimenter's choice in the present moment) determines the effect (the path already taken by the photon in the past).

Resolving the Paradox: Time Travel and Parallel Universes

There is no way to really understand this using any of the conventional interpretations. However, using my idea of synthesizing the transactional interpretation and the parallel universe idea, the paradox is seemingly removed. The photon's quantum offer wave—the wave sent originally—leaves the source and passes through both slits toward the remainder of the experiment. If the screen is up, the photon is absorbed, and the film-emulsion sends back through time a conjugate echo wave that also passes through both slits and is received by the light source. The offer and the echo waves pass through both slits, and the transaction is complete. Here the source wave and echo wave pass through both slits and exist in a single undifferentiated universe. Actually that single universe is a composite of two universes: universe one in which the offer and the echo pass through slit one, and universe two in which they pass through slit two. No attempt is made to tell one from two. So they appear together as a single universe. End of story one.

If the screen is down, the offer wave again passes through both slits toward the telescope; however, now the universes split apart. In each universe only one telescope sends back an echo wave through its corresponding slit. Only one wave is returned in each universe, because of the boundary condition that the wave represents a single photon. And that is a fact recorded in the memory of the observer. It is a fact of consciousness. Someone has observed only one photon. If both telescopes sent back echo waves in a single universe, there would have been two photons present in that universe.

In universe one, telescope one sends back the echo. The observer sees that this is true. In universe two, telescope two sends back the echo, and the observer sees that this is true. In universe one the source receives a signal from telescope one, and the transaction is complete. A similar situation arises at the source in universe two. One photon, two separate universes. An attempt is made to tell one from two. So they appear as two separate universes. End of story two.

Time Is Self-Consistently Connected to Parallel Universes

Wheeler's "choice" also presents us with another new insight. That has to do with the connection of time to parallel universes. If our choices now can affect the past, can we reasonably affirm that choices made in the future affect our present? If so, then we have a new principle appearing in the universe. This is the *principle of self-consistency.* No longer is the universe causal. All that is required is that whatever occurs must be self-consistent.

Self-consistency in this sense means that whatever sequence occurs involving the past and the present, or the present and the future, what happens at both ends of the sequence is logically consistent. Thus if an event in the present causes some event in the past, the present event cannot be one that will send a message back in time that will cause the past to send a message to the present that will cancel out that action. My sending a message back in time to my preteen grandfather that keeps granddaddy from meeting up with grandma would not be self-consistent, since I'm here in the present.

This principle will lead us to another possible connection between parallel universes: the connection provided by information traveling from future to the present as well as from the present to the past. In other words, to the possibility of "time machines."

Chapter
25

BUILDING A TIME MACHINE

In the movie *Dune*, the young princely hero, his father and mother, and others loyal to the family move through spacetime by entering a large cylindrical space above their planet. Floating in space are giant sausage-shaped creatures that are highly intelligent and able to bend space and time. The family appears to go on a mind journey. One moment they are floating above their own planet, and the next they are many light-years away high above the planet Dune, their new home. Time travel is instant, no accelerations, no great expenditure of rocket energy, no reaching the speed of light.

For those of you who wish a more tangible proof of parallel universes than what we can imagine with our own minds, consider the following. If physicist Frank J. Tipler is correct, it could be possible to build a spacetime warping, rapidly rotating cylinder that would enable small objects to time travel to other universes. Tipler first pointed to the possibility of time machines in an April 1974 article[7] that appeared in *Physical Review D,* a volume of physics research explicitly devoted to new frontier edges of physics known as "particles and fields." Tipler's article was entitled "Rotating Cylinders and the Possibility of Global Causality Violation." Tipler, then of the University of Maryland, is now professor of physics and mathematics at Tulane University in New Orleans.

Now as you can surmise, physics articles in journals are a little too obtuse for the nonscientist. What is a global causality violation? It is the existence of a path that winds through space and somehow turns around in time. Whenever something like this occurs, physicists say we are looking at a *pathology*—no pun intended. Now pathologies are rather strange occurrences in physics. They are indications of a bizarre new twist in our understanding. A path that twists through space and somehow arrives back on itself—after traveling through time—at the same time it left is a pathology. So are the exact centers of black holes, according to general relativity without quantum mechanics, where the laws of physics seem to go haywire.

As Tipler explained in the article, classical general relativity does predict pathological behavior. The exact nature of the pathology is, however, debatable. Physicists working in the field have noted that virtually any type of bizarre behavior can be found in the solutions of Einstein's equations. One of these behaviors is global causality violation, and if the usual assumptions about matter, energy, and spacetime are made, then it should be possible to design an experiment that would exhibit this particular pathology.

The pathology indicates that it is possible for an object to travel in the spacetime surrounding a singular region, such as a black hole, so that it returns to its starting point at exactly the same time as it left. In other words, it would follow a path through spacetime that would include a leg where it traveled backward through time so that it could return at exactly the same time as it left. Such a path is called a *closed timelike line* (CTL).

Now a CTL would have to pass twice through a rotating black hole in order to do this. It would, in the language of John A. Wheeler, go down a pair of wormholes and wind up right back in the same universe. This is not recommended, because black holes distort space and time so much that gravity could cause the person traveling this path to be stretched into a stream of atoms as he passed through the hole.

Tipler's experiment involves building a huge rapidly rotating cylinder. The spacetime surrounding the cylinder turns out to be sinusoidally* warped, so that time itself, instead of moving continuously from past to future, varies in an oscillating manner and, if carefully moved through, may not cause such a stretching of material bodies.

Suppose we were, at this point, to imagine that such a cylinder was actually built. Perhaps a scenario such as the following would take place (Just as an aside to the reader, as I created this story, I had to use my own principle of self-consistency several times in the telling. Indeed all story tellers must do so, if the story is to hold together.):

Dateline December 3, 2587: Galaxy News Report

"Tachynauts, a term we are soon to become more aware of, have today brought humanity to the next frontier—time travel. For the first time in history, human beings are about to venture into the past and the future—time periods many

*A sinusoidal movement is one that oscillates and therefore returns to the value it had to begin with. A pendulum swings in a sinusoidal path through space.

A sinusoid is, therefore, an oscillating curve. When plotted as a curve on graph paper, the position of the pendulum bob, as the pendulum swings to and fro, follows a sinusoid as time marches on. If the period of the pendulum is, for example, one second, the bob will return to its starting point every second.

of us believe are either gone or yet to be. In other words, many of us think they do not exist now. If this voyage succeeds, a new vision will become a reality. Three brave persons—two women, Colonels Frances Zweizeit and Carol Raum, and one man, Technical Sergeant Jacob Camino, all from the United Planets Spacetime Service (UPSS)—have today begun what promises to be a journey unsurpassed in human history. A journey into time.

"With us today to help us understand just what is about to occur is Dr. Roland Espacetemp, professor of Temporal Studies at Princeton University.

"First of all, Dr. Espacetemp, can you tell us what a tachynaut is?"

(Professor Espacetemp speaks with a heavy French accent.) "Yes, I'd be happy to. And, by the way, thank you for letting me explain this historical event to the general public. The root word 'tachy' means fast. A long time ago, physicists sought particles that traveled faster than light and called them *tachyons*. Although they never discovered any, the word stuck with us. Thus, a tachynaut is someone who goes faster than the speed of light. Before the age of actual voyages through time, we erroneously called time travelers tachynauts because it was widely believed impossible to time travel unless one was able to exceed the speed of light. However, it turned out that this was not true. It wasn't really necessary to exceed the speed of light to time travel. Light speed was only a barrier in 'flat' spacetimes. In the highly curved world of general relativity, the speed of light was more a nuisance than a real barrier. Hence the word 'tachynaut' stuck in today's vocabulary even though it wasn't necessary for time travelers to exceed the speed of light. When we constructed the time-cylinder, we found a way of getting around the barrier."

"Well, Professor, I am sure that our holo-viewers would appreciate hearing more about the cylinder. Can you give us a little background? When was the time-cylinder first built?"

"Surely. The Tipler cylinder was first completed on June

15, 2583. It was quite an undertaking, and required the services of several planets and a concerted effort and expenditure of galactic funds. Stationed halfway between our own Milky Way and the nearest galaxy in a galactic holding pattern known as a La Grange coordinate or G-5 station (a place where the gravitational pulls exerted by all the galaxies canceled because they pulled in opposite directions), the Tipler cylinder was constructed. It was, in fact, the first human-made cylindrical time machine."

"And today's the day for the first humans to enter it, Professor?"

"Indeed it is now ready for its first test run with human beings. Actually, *entering it* is not quite correct. We hope that none of the tachynauts go too near the cylinder. The gravitational forces are something awful. Instead, they will circle around it at a safe enough distance to avoid being harmed by the gravitational tides generated by the cylinder, but close enough to reap the benefits of time travel in the time distortions surrounding it. By the way, scientists from nearby galaxies, brought together by the United Federation of Galaxies, are still gathering. This promises to be more than just another subluminal space voyage: it is really the beginning of a new era—the age of machined time travel."

"Professor, when did we first conceive that time travel was possible?"

"Frank Tipler, a physicist at the University of Maryland, proposed, more than five hundred years ago, that such a cylinder could be built. But at the time no one really knew how. However, that was before the age of neutron star synthesis—the synthesis of matter found in stars of a certain mass that have undergone gravitational collapse."

"Neutron star synthesis? Would you explain what this is?"

"Yes. There were two major problems in building the cylinder. The first was finding a material dense enough—containing enough matter superpacked together—to create the necessary amount of mass that would set spacetime spinning. The problem was solved by using nuclear matter, the densest substance in the universe. One teaspoonful weighs more than a billion tons. Such matter was to be found only in neutron stars. Since neutron stars were discovered over five hundred years ago, this meant a neutron star roundup."

"But Professor, I am sure that most of our holo-viewers would be amazed at this. Aren't stars huge? How could—"

"Ha, of course, let me explain. Now although neutron stars are quite massive, they are really not very big. They are made from atoms that have been fused together by the powerful force of gravity. Remember, in your early elementary grade classes (*Note*: reader, see Part One), you learned what would happen if the electrons inside atoms were to suddenly fall into their respective nuclei. Of course, this won't happen with ordinary matter because there are insufficient amounts of it wherever life is possible. But in the interiors of dense massive stars, nuclear matter is gathered together by gravity forcing the issue. All atoms are reduced in size. The electrons in these stars become bound to the nucleus, each negatively charged electron fusing with each positively charged proton. When that happens, the fused electron-protons become neutrons—particles with no electrical charge. Neutron stars are made of these kinds of atoms. The nuclei are so close together that the whole star appears to be a gigantic nucleus."

"So did intergalactic scientists search for neutron stars to make the cylinder? Even if they found them, how did they manage that?"

"Believe me it wasn't easy. But we did it with huge capital expenditures and galactic cooperation. We had to form a new intergalactic agency to accomplish this. Remember, we began this venture more than one hundred years ago. And today, galactic nuclear-boys—the equivalents of cowboys in days of old—are quite common. Instead of herding cows, these astronautical 'cowboys' herd neutron stars together using nuclear bomb lariats instead of rope. Find enough stars, fuse them together into a long cylinder, and you are halfway home to building a time machine, wagon-train style."

"Could you elaborate on this, Professor Espacetemp?"

"Yes. This machine required no less than one hundred neutron stars joined together, each about 20 kilometers or around 12½ miles in radius, making a fused neutron star cylinder, 40 kilometers across—about the distance an ancient marathon runner went, and around 4000 kilometers long—about the width of the old United States of America. By making it this long, we were assured that there would be time-reversal zones far enough away from the crushing gravitational tidal forces in the immediate neighborhood of the cylinder's surface.

"The second problem we faced was aligning the spins of the stars so that the whole cylinder rotated with one rotational speed. Typical neutron stars rotate at speeds between 1000 and 10,000 revolutions per second (rps). By herding out the faster rotating stars—all around 10,000 rps, and 'phasing' them so that they rotated as one, the machine was finally created."

"Why did we need to phase them together and why was it necessary to have it rotating at one enormous speed?"

"With this much rotational speed, the outer surface of the cylinder, rotating along the long axis, reached a speed of about three-quarters light speed. This was absolutely necessary in order to get what we were looking for. The time machine required an outer surface speed greater than one-half light speed in order to produce a set of cylindrical time-reversal zones, each zone concentrically surrounding the cylinder. Earlier tests with the cylinder using unmanned spacecrafts containing transmitting robotic intelligences showed us that as they moved radially inward to the cylinder's surface, they passed through alternating time zones, where asymptotic time—the time observed on clocks far from the spacetime warp—ran alternatively forward and then backward in comparison with the local time—as observed by the robots."

"I'm afraid you lost me there, Professor. What is asymptotic time?"

"That's the time we all experience. Everyday time. You see, we are far enough away from the cylinder not to feel the time distortions."

"Professor, I am sure we are all a little lost in understanding time. Can you help us here?"

"It is best and easiest to think of time as already existing, somewhat like space—out there—only in zones much like areas on a map. Let me explain, using the cylinder as an illustration. There are concentric time zones surrounding the cylinder. There are allowed zones where tachynauts can enter safely and also forbidden zones concentrically located—places where time simply ceases to exist. If one tries to

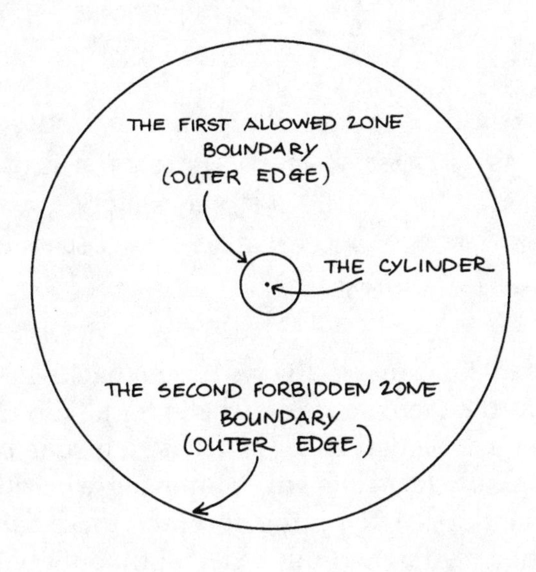

A VIEW OF THE TIPLER CYLINDER SEEN FROM A LONG WAY AWAY. THE CYLINDER IS TWENTY KILOMETERS IN RADIUS. THE FIRST ALLOWED ZONE ENDS AT 700 KILOMETERS FROM THE CENTER. THE SECOND FORBIDDEN ZONE ENDS AT ABOUT 11,500 KILOMETERS FROM THE CENTER. THE CYLINDER IS SEEN END-ON WITH THE AXIS OF ROTATION POINTING OUT OF THE PAPER. IT APPEARS AS A DOT ON THIS SCALE.

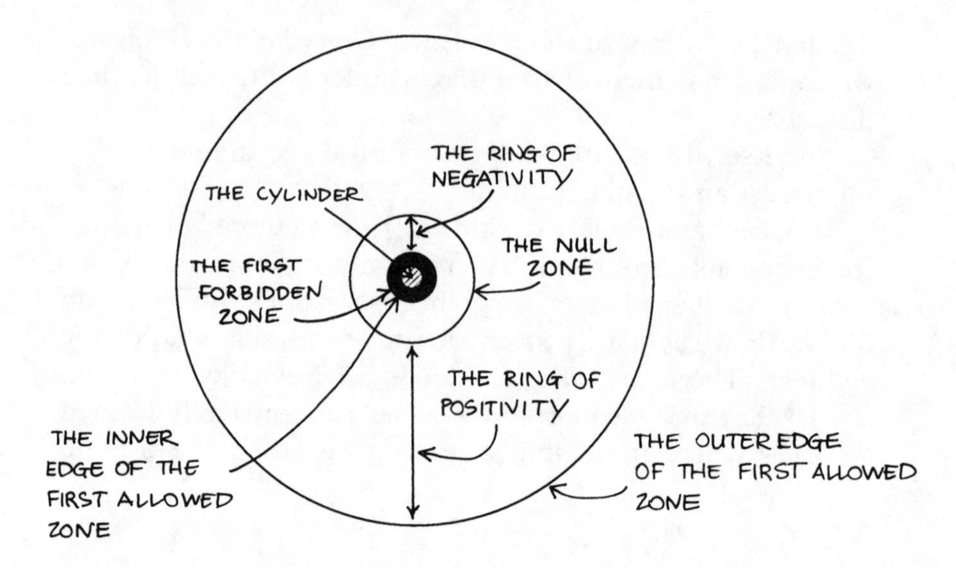

THE CYLINDER IS SEEN END-ON FROM A CLOSER VIEW. IT APPEARS AS A CROSS-HATCHED CIRCLE ON THIS SCALE. THE DIVIDING LINE (THE NULL ZONE) BETWEEN THE TWO RINGS IS SHOWN. THE FIRST FORBIDDEN RING-ZONE SURROUNDING THE CYLINDER IS SHOWN.

enter a forbidden zone, one is always turned back."

"It seems that the tachynauts may be in for a wild trip."

"Indeed. The exact location of the forbidden zone boundaries depends on how fast they will be moving when they approach the boundary. Attempting to enter these forbidden zones will be like trying to climb a steep hill, only with each step they will take, the gravitational pull will become even stronger. Eventually they will have to turn back."

"It sounds like trying to travel through one of those old rotating cylinders—like those in ancient amusement parks. My parents and I went to one when I was a child. Every time we tried to run through the cylinder, it would turn and we would fall over laughing."

"Yes, it is somewhat like that, but I'm afraid it is no laughing matter. The time zones are ever-shifting and dangerous, depending on how fast you are moving. You must be careful. The forbidden zones could cause great damage if you enter them from the wrong direction. In fact, it is not possible to enter the time-travel zones coming in along a radial line perpendicular to the axis of rotation. One must enter stealthily from the side, approaching the cylinder along a line parallel to the axis of rotation."

"Professor, what does the spacetime around the cylinder look like?"

"If we look at the cylinder from an end view, it appears as a 40-kilometer diameter circle floating in space. However, it is surrounded by ever-widening alternating zones of allowed and forbidden travel called von Stockum bands. And—"

"Excuse me for interrupting. Are the von Stockum bands the time zones you have been talking about?"

"Yes, the physicist W. J. von Stockum in 1936 was the first to solve the Einstein field equations for the cylinder. These bands were to time travel what the earlier van Allen belts were to the first earth space travelers. In an allowed zone, time travel is possible and in a forbidden zone no travel is possible.

"The zones are quite visible, by the way. The view is eerie because the boundaries of the alternating zones glow clearly marking the edges of time. The first forbidden 'skin zone' touches the cylinder. This 'skin zone' extends outward 20 kilometers from the surface. This 20-kilometer-thick sausage skin appears as a whirling dervish of colors. Positron–electron pairs burst forward in tiny explosions, and the skin glows a brilliant violet as the matter–antimatter pairs recombine, giving off violet photons before they rain their deadly cascades of gamma radiation signaling their mortal end.

"Immediately surrounding the forbidden sausage zone is the first time-travel 'ring zone.' The ring is composed of two adjacent time zones: the ring of negativity and the ring of positivity. The ring of negativity begins at the skin's edge. Here time runs negatively—backward at the fastest pace. If one could remain there for just a few seconds, it would be possible to travel backward in time to any era. If one moved outward through the ring of negativity, the rate of backward-through-time travel would decrease. The decrease in the negative time rate continues until one reaches the first null zone—the interface between negative and positive time travel. The null zone is extremely narrow and exists only at 130 kilometers from the surface. By remaining in the null zone, one could age as long as one wished, while the rest of the world remained frozen in time."

"I am not sure I follow this. You mean one could enter the null zone and stay for fifty years and then leave it and find the world had not aged one second?"

"Yes, that is just what I mean. Not a very pleasant perspective, is it? A kind of time travel in reverse."

"What will the tachynauts do? They *do* hope to time travel, don't they?"

"Yes, of course. Let me explain. Passing ever outward from the null zone, one passes into the ring of positivity where time begins to run faster and faster forward in time until one reaches the outer edge of the first time travel zone, which is the inner edge of the second forbidden zone.

"This boundary occurs at about 700 kilometers from the cylinder's axis. At the outer edge of the ring of positivity, time runs positively at a near infinite rate. Remaining here they could, in just a few seconds, journey to any event in the future."

"How about the other zones?"

"As one scans radially through the second forbidden zone, one sees how immense it is. Its inner edge is 700 kilometers from the axis, but its outer edge is more than 11,000 kilometers away. The alternating zones grow in size as one moves radially outward, but they also decrease in time-travel strength. Scanning out farther into space, the rings begin to dissolve into nothingness as the demands for normality far from the spacetime warp assert themselves."

"So what will our tachynauts do? How are they going to

use the machine? And how are they going to maintain safety? You've discovered that the forces of gravity are enormous."

"To use the machine, tachynauts will move in along lines parallel to the axis of spin. By choosing their entry trajectory carefully they will be able to enter the first time-travel zone. By maintaining a high enough centrifugal speed around the cylinder they will be able to deal with the crushing gravitational tidal forces. However, their space ships are designed as flatly as possible in order to minimize the crush caused by the difference between the force of gravity at their feet and their heads. In fact, they look very much like the old picture we had of flying saucers—they are simply large, very flattened disks."

"An unusual design, I am sure. Not at all like the spacecrafts of the twentieth century."

"Yes, indeed. Consequently, tachynauts will be lying down in order to time travel. It turns out that the best entry for them will be into the null zone. Here the force of gravity is nearly absent."

"But they won't hang around there too long, will they?"

"No. In the null zone, asymptotic time—the time that ticks far away from the cylinder—stops. The time travelers could enter the null zone and if they wanted to, whirl in donut-shaped orbits about the cylinder. All the while their clocks would show normally elapsed time, but the world outside would remain strangely still."

"What will that look like to us—out here in—ah—asymptotic time?"

"To observers stationed on spaceships at the 'shores of time' the tachynauts' journeys could seem strange indeed. For example, the time travelers could enter the null zone at three o'clock at one location and emerge at exactly the same time from another location in the null-donut. It could be like a grand-scale old-fashioned magic-show."

"Thank you, Professor. I am afraid we are running out of time."

"Thank you."

Dateline December 15, 2587: Galaxy News Report

"Good evening. With us once again is Dr. Roland Espacetemp, Professor of Temporal Studies at Princeton University. I am sure we—that is, all of us viewing—are aware of the terrible tragedy that has occurred. We all feel the sense of loss and extend our sympathies to the family of Sergeant Camino. Frankly, Professor, I am at a loss to explain just what happened and when it happened. Time-travel tragedies aren't the same as space-travel disasters, are they? Can you help us to understand just what did happen and when it happened?"

"Good evening, and thank you for allowing me to explain to our holo-viewers the paradoxical situation that we have just been apprised of. We have just learned that Sergeant Jacob Camino has died—only a few days after his return. He appeared to be hopelessly insane at the time of his death and crying that the world had changed. He kept saying that he hadn't returned to the same world he had left. He was lamenting terribly the loss of his beloved grandfather. However, let me assure you that his grandfather is very much alive and well and, although grieving at the moment, could not understand just what was bothering his grandson when he died."

"Professor, was Camino out of his mind?"

"No, I don't think so. We all feel terrible about the loss. Sergeant Camino was technically responsible for the mission. I'm afraid that the incident was unavoidable. Sergeant Camino violated the *prime paradox*. We never thought any of the tachynauts could violate it. But I am afraid that Sergeant Camino manipulated the vehicle in a way we didn't foresee, and nearly ended the lives of the other two brave tachynauts."

"By the way, Professor, how are Colonels Zweizeit and Raum?"

"They are fine. Just recovering from the shock."

"Professor, I am sure no one knows about the—ah—*prime paradox*. What does it mean? And what happened to the time travelers, particularly Sergeant Camino?"

"Let me get to a few details first. We knew that time travels could possibly cause the tachynauts to enter parallel universes. Indeed, we expected this to happen. We had prepared them for this, and hoped that they would not be foolhardy and attempt any violation of self-consistency. But I'm afraid one of them did."

"Parallel universes? What happened?"

"Once through the null zone, the tachynauts were warned that they would risk all that made sense to them—their inherent sense of time. The tachynauts entered at three o'clock Greenwich time on December 3, 2587, and were instructed to travel back through time a period of a few million years. There and then they were to leave the zone and make holo-observations of the earth from a safe distance. They were briefed that just how fast they would travel depended on the conditions inside the first ring of negativity and how fast they were going at the time of entry.

"In the ring, they circled around the tube, and, like a back-tracking detective, they successfully moved backward through time. By coming close to the null zone boundary, they successfully slowed their passage through negative time and made preparation for the next phase.

"Next, moving away from the cylinder, they began to experience time synchronization as their clocks started to coincide with our clocks in the asymptotic regions far from the cylinder. As they moved to the outer edge, they were expected to travel slowly forward in time. But then Sergeant Camino reversed the procedure and took them to a time zone we didn't expect."

"Why did he do that? Why didn't he follow instructions? Didn't they file flight plans with authorities?"

"Yes. The tachynauts filed flight plans with the Federal

Time-Flight Agency (FTFA). They were supposed to fly a typical sortie taking them into the null zone, then a radial retreat into the ring of negativity to the inner edge. There they were to spend perhaps a few minutes near the inner edge, and then move back toward the null zone. The tachynauts were then to exit, appear in a parallel universe that was younger by a few million years or so, and then reverse the procedure and return."

"Wasn't that dangerous, taking them back so far in time?"

"We didn't think so. It was time recently past that most bothered us, and apparently Sergeant Camino suspected this. Of course there was the consideration of the *prime paradox* (PP) to be taken into account. The 'double-P,' as it is now called, deals with what would happen if the time period they entered did not have a time machine present.

"The paradox was resolved when physicists realized that the only way out was self-consistency. The travelers could only go back in time to a universe where the machine existed."

"You mean that a million or more years ago there were time machines present? How could that be?"

"That was a problem before we discovered parallel universes. In the old days we thought there was only a single universe. The tachynauts could only go to parallel universes where the time cylinder already existed—self-consistently. Even if the past of our universe did not have a cylinder, other universes had them. Thus with the discovery of parallel universes, it became quite apparent that no problem, no PP existed. Indeed, it became quite clear why there had to be parallel universes. They were needed for self-consistency. All events in all universes are webbed together, affecting the whole. Not only did parallel universes provide a necessary base for the existence of matter via the laws of quantum physics, they also provided a means by which self-consistent time-loops could exist. They could exist only in universes where no law of self-consistency was ever broken. In this way, all time travel paradoxes were resolved."

"But what happened?"

"I am getting to it. But let me continue. When we discovered parallel universes, time travel became a reality. A wonderfully consistent picture of the universe emerged. To have any universe it was necessary to have information coming from both the future and from the past. Quantum physics showed that only if both future and past information were present—a *present* was present. In other words, a now-world of physical stuff could manifest only if parallel universes existed. Without parallel universes existing, it was necessary to have an outside agency create a quantum jump or sudden change in the probability of anything. Physicists were bothered by that for a long time."

"But Professor, please tell us what happened."

"I'm sorry. I don't mean to ramble, but it is necessary to have this background to really grasp what happened to Sergeant Camino.

"You see, journeys to the present from the past also fell under the same rule of self-consistency. After leaving the past, the time travelers entered the null zone, moved outward toward the outer edge of the ring of positivity, and experienced a rapid movement through time to the present. They then returned to the null zone and exited into what they thought was this universe where the time machine existed. There they believed they were free to do anything they wanted."

"Anything at all?"

"Not quite. The rules of self-consistency constrained them morally. They would not do anything that would generate a backward-through-time message to their past that would jeopardize them. They believed they carried out their mission without a hitch."

"But it was in the *past* that they faced a problem."

"Yes. There Sergeant Camino violated the rule. And we all saw what happened to him. It seems that when they emerged in what they thought was our present, the computer's memory malfunctioned slightly. Instead of in *this* time period, they emerged about one hundred years ago, but not in our universe. Sergeant Camino realized this before the others, presumably because he was at the control module when this occurred. He noticed that they had emerged in a parallel universe on the day when his granddad was a young fellow and was saved in a boating accident. It seemed that Camino had never really liked the old man—but we didn't know he would ever attempt anything like this. The offender remembered his grandmother's telling him about this day and that his granddad had almost drowned. Sergeant Camino, locating the exact spot using high-resolution holo-loran and a holo-telescope, directed a holo-laser beam at the boat and thereby arranged for his young granddad not to be saved on that fateful day."

"But, if he murdered his grandfather before his grandfather reached sexual maturity, how could Sergeant Camino ever have been born?"

"I am not sure what Camino knew about parallel universes. I think that he believed that if he had killed his grandfather in a parallel universe he would be OK and that somehow this event would alter his relationship with the old man in this one when he returned. He was unfortunately right. Of course, when the others discovered what he had done they managed to subdue him and somehow, with an adjustment to the craft, make it back to our time and this universe."

"But how could he have been born?"

"In the parallel world where his young grandfather never saw an eighth birthday, I assure you there was no Sergeant Camino born. That's where Camino made his mistake. Not being born in a parallel universe had dire consequences for him in those universes where he was born.

"When Camino came back to our universe he noticed that his grandfather had changed. Of course old granddad was still around, but he wasn't the same. Old granddad kept in-sisting that it wasn't he that had changed but that the time traveler wasn't the same. Indeed, everyone who knew Camino seemed to agree with his granddad."

"But—hold on. Did Sergeant Camino ever come back to our universe?"

"The time traveler came back to a slightly different parallel universe."

"This seems impossible to me. I mean, I am here in this universe. I remember Sergeant Camino's leaving us a few days ago."

"It is confusing. However, Sergeant Camino's other universe overlaps with ours in all the quantum superpositions that we all left unresolved. As long as his universe did not conflict with ours, self-consistency was assured. And no one could tell any difference between them. In those regions where his universe differed from ours, it turned out that the differences were mostly felt by him about the universe. All we felt was that he had changed. In other words, he felt as if he had entered a universe that was different from the one he had left. We all felt we were in the same universe.

"As you saw, the direst consequences were felt by him. His other fellow time travelers did not feel this way. They returned to a universe that was perfectly self-consistent for them, the one they had left—just a few days older. The criminal did not. In a few days he seemed to lose his mind. He died shortly after that."

"Thank you again, Professor. This is station KLAS, signing off."

Back to the Present

Of course, this scenario appears strange because we aren't yet equipped with the necessary technology to herd neutron stars and thus build a Tipler cylinder. However, I believe that the rule of quantum self-consistency is real. Parallel universes and time travel do fit together. I cannot find anything wrong with the physics. It also gives us a new understanding of time. In Part Six we will see how this new understanding could make it possible to "talk" to the future. We will look at the work of several physicists who have changed our way of looking at time and mind when parallel universes are taken into account, and we will posit a new definition of reality and existence.

PART SIX

TIME AND MIND IN PARALLEL UNIVERSES

With a bit of a mind slip
You're in for a time slip
and nothing can ever be the same.

"Time Warp"
The Rocky Horror Picture Show

Infinity is just time on an ego trip.
Lily Tomlin

Ego is just time on an infinity trip.
The author

At the beginning of this book and throughout I have suggested that if parallel universes were indeed real, then a profound connection between time and mind would be present. This connection would completely upset our ordinary notion of time. And it seems that by taking the parallel universe theory seriously we will also find that mind enters the physical world in a new and unsuspected manner.

Let me point out that this parallel universe idea is not new. It hit quantum physics in 1957, after all. You old-time science-fiction buffs probably knew about parallel universes before we physicists caught on. I think one of the best science-fiction books dealing with how mind enters a parallel world is Ursula Le Guin's *The Lathe of Heaven.*[1]

The hero in this book is faced with an awful reality: His dreams come true—whatever he dreams becomes real the instant he awakens from his dream. When he awakens he sees a new universe. He realizes, after a while, that he is actually not just dreaming. He is traveling from one parallel universe to another. After each time he dreams, he wakes up in a parallel universe. And everything in that universe is totally consistent with itself. No one else in that universe has any memory of any other universe. The only difference is that the hero remembers the other universe he inhabited before he went to sleep.

Can our minds witness other worlds when we sleep? It just might be that while dreaming, our minds, not being fully occupied with this reality, are able to perceive other realities. Maybe the key to traveling to other universes is to simplify the mind. This world just has too many distractions—such as the art of survival in the concrete jungles of modern life.

Perhaps the unsuspected connections of time and mind to parallel universes are not just a coincidence. These connections indicate that time and mind may indeed be, in some yet to be fully determined manner, the same thing. In this part of the book I want to explore how a few frontier physicists have begun to explore this deeper connection and add a few thoughts of my own.

Some of these physicists see that the problem of measurement—the paradoxical change in the probability of some event once that event occurs or comes to be known—is completely explained by the existence of parallel universes.[2] They then consider how an observer of some physical phenomenon would act if completely alone with the physical object under scrutiny, but subject to the splitting in the universe that occurs with an observation of a physical system. They conclude that if a single observer and a single physical system were totally alone in the universe, the observer would not be able to make up his or her mind once an observation of the system was carried out.

If the system being observed had, for example, two physically possible states, the observer would be of two minds after the observation. It is only when other minds enter the picture that a decision becomes possible.

In a similar manner, if our own minds are simple enough they will only see the superposition of the possible states, but not be able to tell what occurs. As our minds become more complex, the effects of interference from the parallel worlds would diminish, and each integrated complex mind would be able to differentiate between possibilities—one becoming fact and the other fantasy. Perhaps this is what author Le Guin had in mind for her hero.

Thus parallel universe theory may offer some insight into the nature of human intelligence and imagination. It may also help us understand some mental illnesses, such as schizophrenia. Schizophrenics commonly witness other realities —other beings in their presence. They often hear voices, smell things, or have other hallucinations, as we are tempted to call them. But these may not be hallucinations after all. They may be distorted witnessing of other realities, the kind that could appear in parallel worlds.

Although not all of them would necessarily agree with me that this is the case, it appears to me that other physicists' work suggests that with parallel universe theory messages from the future become plausible.[3] We have already looked at one of these physicist's ideas in Part Five. In another interpretation, these messages require intelligence in the present to receive them. An unintelligent receiver would not be able to decipher the messages and thus would make no use of them. As global intelligence increased, these messages would begin to be received by more minds, until the day when the whole planet would be tuned to the future.

Thus parallel worlds imply a new understanding of the nature of mind and time. Another group of physicists is investigating the new nature of time using what are called two-time measurements.[4] Although two-time measurements are quite radical, the inventors of the idea may not believe that it is related to parallel universes. They do not raise the question. They deal with what can be known about a system now if both a past event and a future complementary event (complementary in the sense of the uncertainty principle, like position and momentum of a particle) are determined.

This aspect of parallel universe theory leads to a new concept—the quantum automaton, an elementary memory unit in a quantum computer that records not only the state of some outside observable but also the state of its own memory.[5] This automaton is able to make predictions about its own state that enable it to compute faster than any present Boolean computer element. Perhaps this automaton is at work in our own brains.

The Mind in Parallel Worlds

Is the existence of consciousness—the presence of matter capable of thinking and observing the universe—itself a consequence of the reality of parallel universes? If so, it would be normal to be able to communicate with both the past and the future at the same time. Indeed this may be the only way that we can think or write anything at all. Somehow we must be able to have all that we are about to say or write written somewhere before we "think." Perhaps our brains are time machines—machines that can send and receive messages coming from the past and the future.

If we do receive messages from the future as well as the past, what is the physical mechanism? How does information get around between the parallel universes, and what impact does such information have on our brains? Although the inventors of the two-time measurements theory do not consider this, it appears to me that a means for the operation of our brains arises from two-time measurements—observations of systems that occur at two different times.

These measurements indicate how it is possible to have interference between parallel worlds—not just interference of quantum possibilities as presented in ordinary quantum physics, but interference between histories in the two or more interfering universes.* This consideration has an impact on how it is that human minds can and do make choices. Two-time measurements also show that the future affects the choices we make now just as much as the past does.

Finally, having integrated these ideas as fully as possible, we next consider the nature of reality and existence as seen through the eyes of a parallel universe enthusiast. Consider the conclusion that *reality* includes both the past and the future, but *existence* includes only the present and is totally dependent on the reality of past and future universes. Without them there is no existence now.

*In a recent letter to me, Susan D'Amato stated that her work with Albert and Aharonov did not make specific reference to the "parallel universes" interpretation. She wrote, "While we (at least I) entertain the many-worlds interpretation as an interesting and viable alternative to the collapse postulate, it would be misleading to state or imply that we are 'parallel worlds enthusiasts.'" D'Amato does believe, however, that parallel or possible histories can interfere with each other. She, however, draws the line between histories and universes. The parallel universe interpretation of this is mine.

Chapter 26

MIND IN PARALLEL UNIVERSES

One major problem with the parallel worlds interpretation, indeed with physics in general, is the explanation of our own subjective awareness. How does awareness occur from a physicist's viewpoint? The answer to this, like the differing interpretations of quantum physics itself, varies considerably. Physicists Bryce DeWitt, Hugh Everett, John A. Wheeler, and Neill Graham (DEWG) have no problem explaining some aspects of human memory in terms of computer automata existing in parallel worlds and accept that memory exists in every world where a sequence permits it to exist in a recording device. In some parallel universes, no memory or consciousness, yours or mine, may exist because these universes are statistically too weird to permit this type of device. In other universes, memory devices can exist. Thus memory exists as sequences of events in all but a few parallel universes.

Another slightly more conservative answer comes from astrophysicist Fred Hoyle. He believes that the many universes are simply an overlay of messages from the future. When we become aware of them, we tune to a universe possibility, and thus the other message possibilities are lost forever. In this case, the present universe is cut off from evolving to the future universe whose message was not received. The general tree of all possible parallel universes, with all of its branches, is simply a reference tree that defines the statistical possibilities. To Hoyle, it is the lopping off of the unused branches that makes consciousness possible.

By tuning to a particular future, a particular route and thus a particular parallel universe is chosen. Who is the chooser in this? Hoyle attributes choice to a superintelligence in the future. This superintelligence could be what we mean by God. Or it could just be some new technology in the future that has perfected the means for sending messages back in time from their universe to ours.

Hoyle also points out if all the branches—all other universes—really do exist, then it is possible to have a record in the present that contains memories of other universes, not just the universe we happen to exist in now. Again this may be a hint to abnormal mental states.

Physicist David Z. Albert, from the University of South Carolina and Tel-Aviv University in Israel, has made a parallel worlds model that enables one to "take a photograph" of another parallel world while still inhabiting this one. He also explains how this effect can be used to create a new form of computer—one that can describe itself. This description goes beyond the conventional Copenhagen theories of measurement.

In 1969, Leon Cooper, a Nobel laureate, and Deborah Van Vechten, from Brown University, interpreted the measurement problem of quantum physics in relation to human cognition and attempted to answer the question "How is something known?" in terms of parallel worlds. They believe that a mind simple enough would be able to experience parallel realities at the same time. But a mind as developed as our own could not do this easily.

In the 1980s, David Albert, Yakir Aharonov, and Susan D'Amato pointed out that a "curious new statistical prediction of quantum mechanics" indicated that a measurement in the future can play a role in determining what is known about the present. They then went on to explain how a set of "two-time" measurements (observations of events in the past and future) can create correlations (meaningful relationships existing over several time periods) between observables that would normally not be correlatable. Such observables in the usual quantum theory, as, for example, the momentum and position of a quantum object, are not correlatable because the correlation would violate the uncertainty principle.

Such two-time events sandwich the present moment. This may be the way that messages from both the past and the future interact with the present. Because the past and the future events are determinations of complementary aspects of real physical objects, the messages are each incomplete. I believe that it is only when both messages are received, in the present, that any sense of reality is perceived. The reality we see every day is sensibly possible, in my view, only if we do receive such complementary messages.

Thus, it seems to me that all of these physicists have attempted to bring the mind back into physics using quantum physics as a basis.* So far, these ideas are still too new to have direct experimental consequences for the human mind. However, if computer technology continues to advance as quickly as it has, we will soon be facing computer memory devices that are both macroscopic and quantum mechanical

*However, this is my interpretation of their work and not necessarily their view. This is specifically the case in the work of Albert, Aharonov, and D'Amato.

at the same time. Such progress, as was described at the recent (1986) New York Academy of Sciences meeting, "New Techniques and Ideas in Quantum Measurement Theory,"[6] indicates that even as I write these words some of the questions raised by the new physicists are being answered.

In the next few chapters I want to explore the ideas of Sir Fred Hoyle, of Leon Cooper and Deborah Van Vechten, and of David Albert, Yakir Aharonov and Susan D'Amato. From their work I am able to draw slightly different perspectives regarding parallel universes and the possible role of mind or memory in them.

Quantum Rules of the Game According to Hoyle

Sir Fred, as he has come to be known in knightly circles, a.k.a. Fred Hoyle, Plumian Professor of Astronomy and founder of the Institute of Theoretical Astronomy at Cambridge University, was knighted in 1972 and awarded a Royal Medal by the Queen of England. He is widely known both for his contributions in theoretical physics and astronomy and for his writing of fiction and nonfiction.

One of his novels, *October the First Is Too Late,*[7] contains a real but quite bizarre possibility based on the parallel universes view of quantum physics. As he put it in a later article[8]:

> *There is the possibility of waking each morning beside a different spouse, although our memory each morning will always be consistent with the spouse-of-the-day, and we will therefore be entirely unaware of the other possibilities.*

Hoyle's view begins with his inquiry into the boundary between mind and matter. He points out that it is strange indeed that science has managed to keep consciousness firmly out of any discussions of the material world, in spite of the fact that it is material that writes these words and material that reads them.

Somehow these atoms rattling around inside my brain are aware. We use this atomic awareness—this matter awake—as our consciousness. We think and make observations, and it would seem very surprising that there should be no scientifically mappable interaction between the worlds of mind and matter.

Classical physics has come up from a dive into the pool of deep inquiry empty-headed. Not a clue exists explaining the mind either as a by-product of materiality or as a field of force somehow manipulating matter and bending it to its will.

Quantum mechanics, instead of picturing us as external observers, seems to suggest that we cannot separate ourselves from the events that we are observing. It also implies that our observations may even shape or determine what actually takes place.

In spite of this fact of life, students of science are usually taught that because "macroevents"—that is, events on the scale of human recognition, the so-called everyday life events—involve billions upon billions of atoms, these events are determined by vast numbers of individual quantum events, and therefore depend only on statistical averages which can be calculated with certainty. All quantum weirdness gets washed out in these averages, and the world goes on in statistical mundanity. Large numbers of atoms acting spontaneously make up macroevents and are completely predictable. Examples include the temperature in a room or the pressure inside a water pipe, whereas the quantum microevents that make up these averaged-out macroevents are not predictable.

Hoyle points out this distinction between macroevents and microevents seems quite arbitrary. If we were to interpret this teaching literally, quantum mechanics would imply a spreading vagueness in the world, even to the extent of making everyday life events appear vague. Apparently this does not happen. Somehow there is a precise predictability in the world. Somehow a sharpening of the vague quantum picture occurs. Somehow predictability arises out of unpredictability.

How does this happen? It is not enough to say that an average of unpredictable events produces a macroworld of predictability. Our brains, being collections of billions of cells, are made of billions of billions of atoms, each following a quantum rule to govern its destiny. Thus all atoms in our brains and nervous systems act in accordance with quantum rules. This means that every thought, feeling, emotion, perception of sound, light, color, smell, taste, sexual arousal, altered awareness and normal awareness, every face recognized in a crowd, every melody heard and romanticized about, every love and every hate, is affected by the behavior of the atoms in your brain and nervous system. Thus life with all of its richness as we experience it follows the rules of the quantum game.

Human decisions usually depend on statistical averages of many quantum events, but is every human decision a macroevent made in accordance with these averages?

Hoyle very much doubts it, and I concur. We often make "snap decisions," "irrational choices," make up our mind on the "spur of the moment." While these may indeed only be metaphors pointing to our quantum nature, very few of us have not wrestled with a weighty decision, making up our mind to go one way, and then suddenly, without recourse, completely, "on-the-spot," changed our mind at the moment it was necessary to implement our decision. These snap decisions have the look of quantum events.

Hoyle goes on in his speculations, pointing out how just after quantum uncertainty was discovered, it threw the whole debate about free will and determinism into a new chaos. The Stoics of ancient Greece believed and taught that everything in life was predetermined—nothing came by chance. Descartes in the seventeenth century carried this idea into his philosophy, and it was quite popular even among seventeenth-, eighteenth-, and nineteenth-century physicists. The idea was that for every effect there must have been a prior cause. Now carry this chain of cause and effect back to the first cause and you have the pocket-sized description of causality or determinism.

But then came the quantum, and the antagonists of determinism were refueled. The human brain no longer had to follow the dictates of utter predictability. However, if our brains are quantum instruments, then how does will, free or otherwise, manifest from random events? How can a sequence of heads and tails following the flipping of a coin carry any meaning? After a long sequence of observations of the coin in which a roughly equal amount of heads and tails appears, is there a message somewhere?

Hoyle points out that there indeed may be. In a sequence of one hundred quantum tosses, even though there are about fifty heads and fifty tails appearing, there are about 10^{30} different sequences possible. Perhaps not all of these sequences are random. Perhaps there is information in the sequence itself—a kind of cosmic Morse code. If this is the case, then where is this information coming from? Hoyle postulates that the information comes from the future. Such an idea may at first seem quite weird, but it seems to me that Hoyle isn't the only physicist to postulate such an argument.*

Hoyle believes that only in this way, messages from the future, can quantum events, which cannot be specified by conditions arising in the past, become determined. In this manner, having both a future condition and a past condition, a sequence of intermediate events can carry information—that is, have a meaning.

Hoyle is also bothered by the parallel universe prediction that from a sequence of one hundred tosses of the quantum

*Albert, Aharonov, and D'Amato provide a model that shows how such a sequence could arise, and we will look at their work shortly.

coin, something like 10^{30} different parallel universes spring.

Being quite a storyteller, Hoyle usually attempts to put his ideas in the form of tales. These next two aren't really his stories. I have made them up to illustrate some of his viewpoints regarding parallel universes.

A Knight's Alter Egos

Once upon a time, a knight in shining armor had to face a dragon. Being somewhat reluctant, he decided to flip a coin and see by the choice of the gods of chance whether or not he would fight the beast on that day. So he carried out the experiment and, since this is a fairy tale, flipped a coin. But before he observed the coin and the side of it landing face up, he decided, as most of us would upon anticipating an unsatisfactory turn of the coin, to go "two out of three" and flip the coin again. But this angered the gods of chance and they sent one of their midst, the quantum god, to him. The god then told him, "Because of your fear you must enter upon a journey to strange lands called parallel universes. At first you must enter two parallel universes before you flip the coin a second time."

Not getting what this might mean, the knight agreed—for after all, it should be better to enter upon the god's parallel world denizens, whatever they might be, than fight the fierce, fearful, foul, fiery dragon. Having said this, the god departed and the knight, with some trepidation, observed the coin that had already fallen on the ground. But then, lo and behold, what he saw was not at all what he expected—for with that simple act of observation, he found himself in two parallel universes at the same time.

The knight wondered who was the real knight, me or me? For the knight was cursed. He could sense himself in both worlds at the same time. The knight was cursed with schizophrenia.

So as all stories go, the knight was not at all happy, for he could not make up his schizophrenic mind. In one world he had seen heads and in the other, at the same time, he had seen tails. Did he go out to fight the dragon or did he stay in the castle and contemplate the latest love of his life? Still being in a quandary and without thinking about it any further, he decided to flip the coin a second time. And when it was flipped, the knight and his coin found that each of the previous worlds he had inhabited had split once again to accommodate four possibilities. In other words, there were now four parallel worlds caused by the two interactions of the knight and his coin. Each alter ego of the knight had seen one coin, and each saw one of the four possible sequences of heads and tails: *hh, ht, th,* or *tt.* But this didn't help the knight one bit, because again he was split into all of the worlds.

So the god of chance appeared. And he said, "Hold on, old knight. What seems to be the problem?"

"I can't make up my mind . . . or is it my minds?"

"Well, let me give you a hand. It's your memory that's going bad. Let me do a little rearranging of the sequences you remember. Maybe that will help."

So when the knight was not looking at the coin(s) on the floor the god changed the memory of the knight and the coin in world one from an *hh* to a *th.* And he changed the memory of the knight and the coin in world three from a *th* to an *hh.* And then the god vanished, as gods are often inclined to do when the going gets tough.

Did this help the knight? Not at all, for after the rearranging of his memory and the coin, he was right back to where he had started. He was still in four worlds and all four possibilities of the two flips of the coin were still there. So again, the knight flipped the coin, making three flips altogether. And each of the previously four parallel worlds split once again, making a total of eight. And the knight was even more confused, being now in eight bodies and eight minds.

Again the god appeared and again he rearranged the knight's memory and the coin in each world. But at the end of it all, the knight was still in all eight worlds and all eight sequences of three coin flips were still there.

So he flipped the coin a fourth time, and you know what happened. There were now sixteen knights and sixteen worlds. In each world a knight saw a coin lying on the floor of the castle. And in each world the knight believed he remembered how the coin had previously landed. But his memory was playing tricks on him because the god had once more rearranged the coins and his memory in each parallel universe, and the tricks didn't change anything. The knight being even more confused decided to flip a fifth time.

And as all stories go, the gods of chance intervened once again. And said to each alter ego–knight the same thing: "We are bored with this display of cowardice. One of you must go out and face the dragon." But the knight(s) quivered in the castle corner and started to flip the coin again. But before he (they?) did, a dragon entered the castle (actually sixteen castles) and ate the knight—first roasting him with his breath, of course, since knights are a bit stringy when eaten raw. End of story.

But suppose we stop the story at the fourth flip for just a moment. Let's look at what the implications are. Normally, we aren't aware of parallel worlds. If the gods of chance hadn't entered the story, neither would the knight be aware of his alter egos. In each universe there would have sat a knight-observer, an alter ego of the first observer. Without the intervention of the god, each alter ego would have believed that he was the only ego around and what he saw was the same coin that he started with. In other words, what he saw previous to the next flip was the coin laying on the floor with a particular side showing. And what he thought he remembered was the sequence of heads and tails he had previously flipped.

With each flip he wouldn't have felt the universe split at all. He wouldn't have realized that after the fourth flip there existed fifteen other universes besides his own. In his mind, so he would have thought, a sequence of four observations resided in his one and only memory. And that sequence must have all been in his universe, the only universe there was.

However, this wasn't quite what happened, was it? The quantum god of chance had entered. And he did play some mischief. When the knight-observer was not looking, the god jostled the coins in all of the universes, so that at the end of the jostling, the previous sides facing up were no longer the same in each universe. For example, in world twelve the god changed the third head in the sequence to a tail, and in world three he changed the third tail to a head, just to keep everything in balance.

Now, it turns out that so long as everything is in balance, there is no way to keep track of what the gods of chance are doing. The only difference between the story and real life is that we don't normally know we inhabit parallel universes. Such jostling is perfectly possible in the quantum world and can be described by the standard rules of quantum physics. At the end of the jostling in the real world, each coin has evolved so that regardless of how the coin landed on the floor there is equal chance once again for finding heads or tails.

There is no requirement in the real world or in the knight's world, because of the quantum gods of chance playing their games, that the alter ego in any universe remember the original sequence he started with. His memory bank may contain the memory of another alter ego. There is no requirement that the observer of the first coin flip be the same observer of the second. All that is required is the consistency of four observations.

In the story, there was no requirement for the state of consciousness of the knight following the second flip to evolve from the state of consciousness following the first flip. And in real life, according to the many-worlds interpretation, in the second periods of our lives we may be remembering the experiences of an alter ego in the first period. In other words, what is in our memories may be residues of lives that we have never actually lived. The second story has us consider this possibility.

The Spouse-of-the-Day

It is here where Hoyle's amusing quote about the spouse-of-the-day can be illuminated. In a tiny village in New Hampshire, Mr. Jones goes to sleep next to Mrs. Jones, one quiet Sunday evening. Nothing unusual about that. Mrs. Jones smiles at her husband sleepily, and remembers the violets he brought home for her birthday—something he rarely manages to do because the poor dear works so hard at the office. She really thinks he spends too much time at the office these days and wonders if he really should. His late-night meetings are a bit of a strain for her. But Friday last he did remember. She is really quite pleased that he brought the flowers, even though she has hay fever—something he never can seem to remember even after fourteen years of marriage—or is it thirteen? She falls asleep and dreams of knights carrying her from one castle to the next.

When she wakes up next morning there lies Mr. Jones. Alongside of the bed she sees the familiar shine of his heraldry brimming with reflected light from his shield. The familiar sounds of the peasants working in the fields fill her ears, and the smell of horses from the king's stable reaches her delicate and sensitive nose. She awakens with a start and immediately begins one of her sneezing attacks.

Mr. Jones wakes up surprised to be arising so early on a Monday morning after chasing bandits out of the castle stronghold last night. His wife smiles at him, "Sor-ree. But you do know how flowers make me sneeze. They are such lovely violets, but, dear, pay a little more attention to me and try not to spend so much time at the round table drinking beer and listening to that old fuddy tell such stories."

Mr. Jones—Sir Lancelot Jones, Esq., to be more exact—looks down at Lady Gwen Jones wondering if this was the same woman he married thirteen years ago.

The above episode implies that there are really two

Joneses. One couple lives in the present and the other in the past. They are really split off from the same Jones's egos.

Now I've carried the story a bit far by placing the alter egos in two different time periods. I'm sure that wasn't Hoyle's original concern. He, I'm sure, would have had both the Joneses living in the same time period. However, time is a bit of an embarrassment in quantum physics. We don't exactly know just what to do with it.

Perhaps the future Joneses are in communication with the past Joneses. In fact, this may be how the mind or memory works. All time exists now, and memories or anticipations are messages from parallel worlds that exist in the future and the past, but at the same time all lying side-by-side in the present.

Chapter 27

QUANTUM "TWO-TIMERS" AND MORE MESSAGES FROM THE FUTURE

The major paradox of parallel universes is that it is hard to believe that at this moment there are an infinite number of other *me*s in the world, with a new *me* cropping up every time I decide to do something rather than something else. In fact, one way out of this parallel universe paradox is to take Hoyle's idea of messages from the future seriously. If the sequences of our lives are being dictated by information coming from both the past and the future, it seems that only one of the parallel universes needs to be inhabited at any given time—the one that we are subjectively aware of. The other universes pass us by because we don't pay enough attention to them to have them register on our nervous systems and brains.

The messages from the future and the memories of the past interact in this universe in a rather interesting way, producing a modified memory that appears to have well-determined records and concepts. By well-determined, I mean that there appears to be no uncertainty present in our brains in violation of the uncertainty principle.

In this new memory, which also contains some anticipation of the future that has already existed and is sending messages back in time to us, the world appears to be quite determined. We know where we've been and we have a distinct sense of where we are going. Both the positions and the paths of our brain particles appear to be known, and yet we still seem to not quite be in control of it all.

How could such a situation arise? A surprising answer has come from three physicists working together in South Carolina. David Albert, Yakir Aharonov, and Susan D'Amato, in a series of papers[9] published in physics journals over the past few years, have noticed a "curious new prediction of quantum mechanics." It is possible to have knowledge of both the position and the momentum of a particle at the same time in violation of the uncertainty principle and yet not violate the uncertainty principle or any other quantum mechanical law.

How to Violate the Uncertainty Principle: Talk to Yourself in the Future

What is at issue in their calculations is the time period between two determinations. Albert, Aharonov, and D'Amato (AAD) point out that if a measurement is made of the position of a particle in the past, and another measurement of the momentum of the particle is made in the future, then both the position and the momentum of the particle are knowable with certainty in the present. Now before anyone points out that this obviously violates the uncertainty principle, one needs to consider that in the present one hasn't actually made the momentum measurement yet. Thus, as far as we can practically determine, the momentum is still uncertain. It is only when you look at the situation from a vantage point beyond the time stream that you see AAD's point of view.

This viewpoint is strengthened by Hoyle's speculation of future messages. It also provides a convenient way to bring order and determinism back into the universe. The order is, however, not in violation of the uncertainty principle as we experience it, because it turns out this principle only works for processes going one way in time—either from the past to the future or from the future to the past. It fails if both future-to-past and past-to-future processing is allowed.

However, this doesn't get rid of choice. We still must make choices based on the information coming to us from both the past and the future. The choices we make determine which of the many paths through the parallel universe maze we are to follow.

And it also presents a new paradox—the paradox of time. If the future does send messages back to us, then doesn't that imply the existence of a future, now? We take it for granted that the past was real. We see it in our memories. However, if that was all there was to memory, our minds would be bombarded by the vagaries of the uncertainty principle and we wouldn't be able to remember anything at all. Using the AAD theory as my guide, I think that the ability to remember the past is based on our ability to remember the future as well. Both the past and the future must somehow have previously existed as far as our memories are concerned. Only in this way can any determined record exist.

Thus any memory of the past—something we witness in the present—is also being determined by the future. But if the future exists, just where is it? In a parallel universe, naturally.

Chapter
28

TAKING A PHOTOGRAPH OF ANOTHER PARALLEL UNIVERSE

The following thought experiment is probably the most outrageous idea that one could entertain, if one is firmly convinced of the existence of this world and, of course, no others. Physicist David Z. Albert, the originator of the idea, believes that it is only a matter of time before a parallel world will be detected. To be perfectly fair to Albert, he doesn't really like the name *parallel universes.* As he put it[10]:

> *It is probably too bad that this very simple thesis has come to be called the many-worlds* [or parallel universes] *interpretation at all, because that name has sometimes given rise to the false impression that there are supposed to be more physical universes around after a measurement than there were before it. It might have been better to call what Everett came up with a*

> *"many points of view" interpretation.... However, the rules of Everett's game, which he insists we play out to the very end, require that every one of the physical systems of which the universe is composed (including cats, measuring instruments, my friend's brain, and my own brain) can be, and often are, in those bizarre superpositions. The various elements of such a superposition, in the case of brains, correspond to a variety of mutually exclusive points of view about the world, as it were, all of which are simultaneously associated with one and the same observer.*

Thus while Albert doesn't feel that there is more than one universe, there are an infinite number of mutually exclusive viewpoints about it. And even more, these viewpoints can exist simultaneously in one brain. Here lies a clue to the existence of multiple personalities. It would certainly be exciting to be able to detect that other "viewpoint" while we still held our own. Perhaps this is what a multiple personality does.

Get Your Cameras Ready

Now we can look at Albert's explanation of taking a photograph of another parallel universe while we are still in this one. To grasp the idea clearly, consider once again the example of the tiny molecular-sized quantum coin, discussed earlier.* The coin can exist in either a side-showing state (with heads or tails showing) or in a color state (with red or green appearing). Color is complementary to side-showing. You can't measure both simultaneously. This means that because of the quantum rule of complementarity, you cannot determine both the color of the coin and its side at the same time without violating the uncertainty principle.

*See Part Two, Chapter 8.

The Rules of Parallel Universe Photography

Let me review these rules by using the following example. You have a new special type of camera. It can detect light at various wavelengths. Long wavelength light causes the film in the camera to register the color of the light. But short wavelength light has no color; when it enters the camera, the film is able to register very fine details of the object that the light has reflected from, such as a tiny engraving on the small molecular-sized coin, but no color is registered.

You can set the camera in either mode. If you decide to observe the color of the coin, red or green, you will not be able to see which side of the coin the light reflected from is facing up. If you decide to look at the side of the coin, you will not be able to determine the color of the coin. This is because the experimental device you are using to make either observation precludes the other.

To see a color you must use white light and the wavelengths of white light are too long to discern the fine details of the carved surface of the coin. If you look at the side of the coin, you must use much shorter wavelength light, so short in fact that no visible color is discernible.

In the parallel worlds model, the universe in which the observation of the color red or green occurs is actually a superposition or overlay of universes in which the sides heads and tails are observed simultaneously. Similarly, the universe in which the side of the coin, heads or tails, is observed is also a superposition of universes in which the colors are seen separately.

Taking the Photograph of a Parallel World

Suppose now that two friends, Jack and Fred, carry out a set of observations. Suppose that Fred decides to use the camera and observe the side-showing state of the quantum coin. Accordingly, when he sees the coin he and the coin split into two parallel worlds. Fred sees heads in one world and tails in the other. As far as he is concerned, in each world he sees a definite side of the coin showing and is not in any position to predict the color of the coin. Yet both worlds really exist simultaneously.

That is, the superposition of worlds in which Fred sees heads and the coin has heads showing and the world where Fred sees tails and the coin shows tails is itself a world capable of being measured by another observer who can photograph and thereby observe the superposition of the two universes. The second observer will use the same type of camera to see the superposition of the two side-showing parallel universes. Thus he will photograph a color.

Suppose we call this world where Fred, his camera, and the coin are in two side-showing worlds "the color world." It corresponds to a state of both Fred and the coin together, even though Fred does not see any color at all in either world that he inhabits.

Now suppose that a second observer, Jack, comes along and no attempt is made by Jack to photograph what Fred sees. What Jack is interested in is taking a colorful picture of Fred and the coin. Jack is not interested in the coin's sides. So Jack uses white light. Now Jack doesn't know which color Fred and the coin will reflect. It's a fifty-fifty chance that Jack will photograph a red or a green color for Fred and the coin. Perhaps Jack sees Fred and the coin together reflecting red. Of course, as far as Fred himself is concerned, no color is apparent at all. He doesn't see the color picture because he is split by his own observation of the coin. All Fred sees is heads or tails.

But now suppose that Jack takes the picture and shows Fred what Jack observes. He tells Fred that he sees him and the coin in a beautiful red light. He even shows Fred a photograph of Fred and the coin glowing a cozy red. Fred looks at the photograph and is amazed, because from his own point of view he doesn't know the color state of himself and the coin. And even more amazing, Fred realizes from the quantum rules that if he and the coin are reflecting red light, then another Fred must exist—the one in the parallel universe.

As soon as Fred sees the photograph his memory is also altered. It now contains two records. In universe one he sees the photo of himself and the coin glowing red, and he believes that the coin is showing heads up and the coin is showing heads up. In universe two he sees the photograph of himself and the coin glowing red, and he believes that the coin is showing tails and the coin is tails up.

This state of things is weird because Fred is confronted with two pieces of evidence that at first sight seem to violate the uncertainty principle. First of all, Fred is in position to make a prediction about the glow state. He can predict that he and the coin will be glowing red even though he sees that the coin has heads showing (in universe one, that is). Thus while inhabiting universe one he is aware of the existence of universe two because the red glow state can only be observed if a superposition of both side-showing states of the coin exist at the same time. Thus he now knows that the other world exists because of what his buddy has shown him, even though he is not directly aware of it.

Furthermore, Fred is now in a position to predict with certainty that if he were to photograph the glow state of the coin and himself it would come out glowing red—a measurement which proves the existence of a world other than the one he presently inhabits. While having a photograph of the coin in his hands that shows the coin has heads up, he is able to make this remarkable prediction. He can predict with certainty that the coin and himself under white light will reflect a red color. If he were to photograph the color and then the side of the coin (in that order), he would be able to predict both results.

However, if he were to repeat his observation and photograph the coin's side first—in other words, confirm the side of the coin he sees—and then use white light to see the color of the coin, he would not be able to predict the color because the measurement would involve only the one universe he was inhabiting. If after making the side-showing measurement, Jack were to come along once again and measure the color, Jack could observe it to be either green or red. Thus the order of the measurements is important.

This tactic of looking at photographs of oneself inhabiting a multiplicity of parallel universes is a strange business that may have some bearing on certain mental states of human beings. It could involve a means for understanding some forms of mental disorders, such as schizophrenia, in which a person believes in the existence of events outside of the universe he inhabits.

I remember a strange conversation I had with a diagnosed dangerous schizophrenic a few years ago. I will call him Ron, although that wasn't his real name. He was troubled by what he perceived to be the presence of a strange man who followed him wherever he went. This man he believed to be God or a messenger from God. And this man told Ron to kill his wife. In a very ghastly murder scene, Ron did just that and was soon locked up in the nearest asylum. The story he told was frightening, particularly since Ron seemed so sure that the "man" was real.

When Ron told me this story, he was already highly subdued by a powerful drug that kept him pretty well locked into the present. Even though the "man" no longer appeared to him, Ron still had to be under constant care, and he remembered the scenes of his actions with great clarity, even though the events had taken place more than ten years before I talked with him.

Perhaps Ron was seeing himself in another world. Perhaps his sick brain was capable of "photographing" his alter ego. The tragedy was that this ego was perceived as sickly as Ron's mind was capable. Is there such a thing as a good schizophrenic?

Although the application of parallel world photography to mental disorders may have to wait for a while, it may not be long before the concept is used in the design of a new form of computer—a quantum computer. Here, a series of parallel processes takes place in one memory location of the computer and then at the end of the processing the superposition of the results are put back into the memory. That memory state is then later looked at. Such a scheme[11] was first mentioned by David Deutsch of Oxford University as a means for solving a class of problems that cannot be solved in any other way.

Using Parallel Universes to Predict the Stock Market

Deutsch imagines a very practical use of his idea—predicting the stock market. He has conceptualized a two-part investment computer program that enables one to estimate the next day's stock exchange movements given today's. The program contains an investment strategy based on the results from the previous day. However, the problem is quite complex and requires a running time of two full days—one full day for each part of the program. Since each part takes a full day to run, the program is quite useless. By the time the computer had finished its calculations, the day for the investment would have passed. Tuesday's stock predictions are not very useful on Wednesday.

Deutsch says, however, suppose that both parts of the program are run in one memory location in parallel worlds on the same day. Then the program would be completed in time for the next day's market. However, there is a problem with running programs in parallel universes. Since they are parallel programs, the result will not always be accurate. In fact, in only one day out of two on average will the prediction of the program be accurate. On that day, a successful investment can be made. On the other day, no investment is made.

To see why this is so, we need to return to the quantum coin example. Remember that when Jack enters the picture, he shows Fred that Fred and the coin are glowing red even though Fred in each parallel universe only sees a head or tail. Since it is also possible for Jack to look at the superposition of Fred and the coin and see the color green, Jack could have shown Fred that he and the coin were glowing green. Jack can't predict ahead of time which color he will see. There is a fifty-fifty chance of catching Fred and the coin in a red or green state.

Computing a successful strategy is like that. Fred seeing the side of the coin showing heads is part one of the program. Fred seeing the side of the coin showing tails is part two of the program. Both sides must be seen by Fred in order that the program succeed. But the combination witnessed by Jack could be red or green.

Suppose that the red state corresponds to the program successfully completing a strategy and the green state corresponds to a failure. When the strategy is correctly computed, a memory bit shows the color red, whereas if it is not successfully computed, it shows the color green.

Thus the practical investor has a distinct advantage of *weighting his bets* to invest when and only when the computer bit shows red in its memory. On that day a successful calculation of the strategy is made.

Deutsch believes that quantum computers will be possible in the near future. He believes that they will use magnetic flux quanta as the fundamental logic units—instead of today's on-off fixed Boolean logic elements—because they can be superimposed in the quantum physical sense. Professor Deutsch believes that the Everett parallel universe model is not just a question of interpretation, but a testable reality. He points out, as I have stated in my previous book *Star Wave,*[12] that a true artificially intelligent computing machine will not be realized until the kinds of quantum interference effects, such as those the parallel universe theory predicts, can be measured.

Just how ethical it will be to use a machine, no one can say. If this computer can see into the future, it would drastically alter the course of human events, no doubt. In Chapter 30 we will consider some of the ethical problems of parallel universe computers.

Chapter 29

HOW IS SOMETHING KNOWN?

When we say we know something, what do we mean? We usually mean that somewhere along the ways of our lives we have had an experience. Perhaps we weren't so sure of this experience, so we checked it out with our friends or our parents, spouses, etc. "Oh, yeah, I remember when that happened to me," our friend might tell us. "Nothing to worry about." And we walk away with some comfort that the experience wasn't really out of the ordinary.

Most often it is a friend that reassures us that our knowledge is OK, that we have correct information. Here we want to explore a little farther just how mutual agreement "creates" reality. Should I say, "mutually assured reality?" Is our detente path of "Mutually Assured Destruction" (MAD) our own reality creation? Perhaps the reality we have come to accept is not really as simple and personal as we might think. What we see, hear, smell, feel, and taste might very well be the senses we have because we have come to agree with each other just what those senses are supposed to sense. According to some quantum physicists, this may indeed be true.

The chapter title question was first given a quantum physical answer[13] by physicists Leon N. Cooper and Deborah Van Vechten of Brown University in 1969. Cooper and Van Vechten (CVV) interpreted the problem of knowledge as a problem involving the measurement of a physical attribute. Whenever a measurement of a physical system takes place, the system appears to "jump" into one of many possible physical states. According to quantum physics, the system cannot exist in any of these states before the measurement occurs. So far, there is no way to account for this jump without using some additional concepts.

In other words, a person comes to know something if he or she comes to measure or evaluate something. The measurement problem is unique to quantum mechanics, but is a mystery, if one does not posit the existence of parallel worlds. Accordingly, whenever a measurement takes place, the universe splits into as many universes as there are possibilities.

However, one physicist, Evan Harris Walker, from the Aberdeen Proving Ground in Maryland, believed that there had to be an experimental way to decide the issue.[14] Not liking the parallel universes theory, he opted for adding "consciousness" to quantum theory in a different way. His solution was that operating underneath quantum physics was a set of hidden variables that "pop" the question and force the system to assume a value when the system is observed by a conscious being.

Thus to Walker, consciousness represented a hidden variable—one that was not under normal control because few of us can control our minds or thoughts to the power necessary to make the control appear. However, he suggested certain experiments which would exhibit "psychic" forces or minds influencing matter.

To Walker, who was upset by splits in our universe every time something came to be known, physicist Bryce DeWitt, the strong advocate of parallel universes theory, pointed out the distress most of us felt when we were confronted with the splitting of the entire world for the first time.[15] In Walker's model (which avoided the split), consciousness acted to make the choice, so that the branch we observed was the only branch there was. DeWitt explained that this was not the only way to add consciousness to the theory. He referred to the work of Nobel laureate Leon Cooper and his associate, Deborah Van Vechten, from Brown University, who provided a definition of consciousness quite adequate for physics.

The theory of consciousness put forward by Cooper and Van Vechten is perfectly consistent with the parallel universe theory of quantum physics. Instead of a measurement suddenly producing a single answer, the parallel worlds view says that each possible answer is produced in a parallel world.

CVV considered this problem as one completely solvable by the existence of parallel universes. There would be no need for other worlds if there was some better way to explain how a quantum superposition of contradictory results suddenly appears to reduce to just one of its values when an observation takes place. So far, no one knows how this happens.

CVV, apparently working independently of Hugh Everett,* even though Everett's work was well known in 1969, followed a line of reasoning similar to Everett's. They ask us to consider how our minds come to know a fact. This seemingly obvious ability to know facts—that each of us is capable of making up our own mind as to what actually takes place in our environment—is difficult to explain when quantum physics is brought into the picture.†

For example, suppose we consider the simple double-slit

*The inventor of the parallel universes model of quantum physics.

†It is even more difficult in classical physics, I believe.

experiment once again but with a slight change. Here a single subatomic particle is allowed to pass through space from a source to a screen. Before it encounters the screen it must pass through a barrier that is opaque except for two slit openings, one above the other. The particle must therefore pass through one slit or the other if it is to reach the screen. Thus after passing through the barrier the particle has either passed through the *lower* slit and been detected there, or passed through the *upper* slit and been detected there, according to common sense.

Now, according to quantum physics, both possibilities must coexist–hence the idea of two parallel universe particles. Yet only one particle is "really" present, as we shall see. After passing through the slits, the particle lands on the screen and is recorded. An observer comes over to the screen and notes where the particle has landed.

However, it is possible to conduct another experiment that actually detects that the particle has passed through both the *upper* slit and the *lower* slit. Thus a single particle simultaneously follows two paths toward the screen. Using a magnetic field arrangement, it is possible to add* the two paths together, allowing them to interfere with each other. This interference shows the existence of both paths simultaneously. This is accomplished using what is known as a Stern-Gerlach measuring technique. I won't explain the details of the technique because it is not germane to the argument.†

This is nothing more than the wave-particle duality described earlier. When the particle passes both through the *upper* slit and through the *lower* slit, it acts as a wave. When it passes through either slit, it acts as a particle.

CVV argue that as a wave, even though the particle eventually is recorded at only one point on the screen, the particle

*This is a superposition of the two paths.

†Although the details are different, it is quite the same principle as I described in the previous chapter for the quantum coin. Detecting the coin's color is the detection of the superposition of the coin's head and tails states.

must nevertheless pass through both slits. To make their point even more apparent, suppose that two separate detectors are employed, each situated just behind its respective slit. When a particle is detected in either detector, the detector will record a reading. If the *upper* detector records the particle passing through the slit, it will read: *particle observed by detector in upper slit.* If the *lower* detector records the particle passing through the slit, it will read: *particle observed by detector in lower slit.* Before the particle passes through, the detectors are both in the empty—no information state. After it passes through the slits, the particle and the detectors are in parallel universes. The upper detector has detected a particle so that it records *particle observed by detector in upper slit,* and the lower detector records *particle observed by detector in lower slit.* In the upper detector we have the "fact" that the particle has passed through and is thus an *upper* slit particle, and the detector has changed from just empty to *particle observed by detector in upper slit.* Together their state is *upper particle* AND *particle observed by detector in upper slit.* This product is the same as when two possibilities are multiplied together. A similar product is produced for the lower slit: *lower particle* AND *particle observed by detector in lower slit.*

This story is somewhat like that of the man who met a man while going to St. Ives. First he meets the man who has seven wives and each wife has seven cats and each cat has caught seven rats and so on. How many are going to St. Ives? Thus we have both *lower particle* AND *particle observed by detector in lower slit* AND *upper particle* AND *particle observed by detector in upper slit* states adding together. Now comes an observer named Mike whose mind state is temporarily blank. After seeing the results of the experiment his mind splits into two states as well: (1) *Mike observed upper detector particle,* standing for his mind observed the *upper* detector has recorded that the particle has passed through the *upper* detector and (2) *Mike observed lower detector particle,* standing for his mind observed the *lower* detector has recorded that the particle has passed through the *lower* detector. Thus the state of the mind of Mike the observer and the detector and the particle is *lower particle* AND *particle observed by detector in lower slit* AND *Mike observed lower detector particle* plus *upper particle* AND *particle observed by detector in upper slit* AND *Mike observed upper detector particle.*

This is just a long way around to say that the universe has been split into two parts. In universe one the particle has passed through the lower detector, the detector has detected the particle in the lower detector, and the mind has seen a particle in the lower detector. Now repeat this for the upper detector.

Thus it would appear that our observer, Mike, cannot make up his own mind. He is of two minds, yet somehow he is of only one mind recording the particle has passed through the *lower* detector or the particle has passed through the *upper* detector, as fact, but not both. If some other observer, Alan, comes along at this time, and somehow can get into the minds of the observer and the detector and the particle and bring them back together, we would have a situation somewhat like the one mentioned in the previous chapter—a photograph of a parallel world. One can't do this, because the mind doesn't allow itself to be measured like a particle. But according to the parallel worlds idea, even though both mind sets are present in Mike's brain, he doesn't know it. CVV believe that if his mind were simple enough, he *could* know it.

How is this possible? CVV explain that we alone cannot make up our minds. We need additional information coming from other observers that concur with us that our observation is the right one. Suppose we add some other observers to the picture. Let them have empty mind states originally. Call them Alan, Barbara, Charles, Diane, etc. After observing the detector their minds will be in the following possible states: *Alan has observed upper detector particle, Alan has observed lower detector particle, Barbara has observed upper detector particle, Barbara has observed lower detector particle,* and so on, corresponding to each observer having made note of which slit the particle passed through. Thus the state of the world of observers, the slits, the detectors, and the particle is

> *WORLD ONE—lower particle* AND *particle observed by detector in lower slit* AND *Mike observed lower detector particle* AND *Alan has observed lower detector particle* AND *Barbara has observed lower detector particle* AND . . .

plus

> *WORLD TWO—upper particle* AND *particle observed by detector in upper slit* AND *Mike observed upper detector particle* AND *Alan has observed upper detector particle* AND *Barbara has observed upper detector particle* AND . . .

In each branch or universe there exists a world of agreement. And the matrix of minds, being made as complex as possible by having more minds in the fold, is composed in such a way that there is no longer any interference possible between the two branches. This is where the buck stops for CVV. Once all these minds have entered the picture, the possibility for interference vanishes, and the worlds are completely split apart. Once this happens, each world coexists without any further interaction. From this point onward each separate world is a real world made up of knowing minds and there is no confusion possible.

I can't help but think of those who suffer from *multiple personality* syndrome here. Perhaps each person's mind is made from many different personalities. If they all agree, there is no problem. The Alan, Barbara, Mike, Diane, and Charles inside of me have no basic disagreements. We all agree that we are Fred. But in a battered and abused child, this might not be the case at all. Alan and Barbara do not agree what to do when little Freddie gets knocked in the head by his daddy. Freddie goes off to his room and must sit alone. Does Freddie split? Does Freddie's mind come apart into

Alan and Barbara disagreeing about just what might have occurred moments ago when daddy took out his strap?

A disagreement could cause Fred's head to take on either Alan's personality or Barbara's, because they no longer agree what is to be done about the deplorable situation they find themselves in. Here we find interference between the two or more worlds. When Fred splinters into Alan, Barbara can no longer be heard or tolerated because she disagrees with Alan. When Fred becomes Barbara, Alan's presence can no longer be felt. Is this the solution to multiple personality problems?

Perhaps the reality we, who are "normal," perceive could be much different if we were to change our viewpoints about the world, not agreeing with the majority on all things. The world we inhabit branches out from the many, continues to split off, as we become more in agreement with each other. What emerges in each world is a stagnant society rife with rigorous rules.

Other groups or nations might not see the world as we do. While there may be much that we can agree on, there may be things we cannot because we don't live on the same planet, even though we walk the same earth. Perhaps with a parallel universes perspective, we can learn to be more tolerant of each other. After all, who really has the only reality there is? Even though there is only one going to St. Ives, all worlds, nevertheless, are still there.

Chapter
30

QUANTUM COMPUTERS AND QUANTUM ETHICS

According to physicists Cooper and Van Vechten, until the many minds entered the picture, schizophrenia in any single mind was entirely possible, and indeed maybe even necessary. If a mind was simple enough, it would be possible to merge its two parallel branches into one using a technique similar to the Stern-Gerlach magnetic technique. How this would be done in a human brain is unknown. But a computer memory system is another matter. There, magnetic fields abound and are necessary to record or detect* any information.

It certainly seems possible that a new generation of quantum computers will be invented in the near future which will use this "simpleminded" quantum physical superposition technique to an advantage. One can imagine two parallel

*Write or read in the computer jargon.

streams of data processing both taking place in a single memory or processing unit which uses superconducting currents of electricity as detectors. By looking at the merger of the streams or currents after a series of parallel processes, a new state will emerge that is impossible with contemporary computers. This new state will contain a memory of events that never occurred in any one universe but could only be produced by quantum physically superpositioning the currents and thus having them interfere with each other. Until that day, however, we will need to watch and wait.

Quantum Ethics

If quantum computers do become a reality, some new questions of ethical standards will arise. If we find that our observations of facts depend only on what we are in agreement about, then the past cannot exist as a series of indisputable facts. Thus, using CVV reasoning, the past is a question of many minds coming to agree that *that* was the past and no other. Any single mind would be helpless to the onslaught of interfering parallel universes. We need each other to provide a basis of reality.

However, therein lies the rub. Many groups form that indeed reconstruct the past to fit their interpretation. The neo-Nazi movement comes to mind. By reinforcing the picture that few people died in Nazi death camps, perhaps 60,000 instead of 6,000,000, this movement has created a past that, if enough of us believe it, will indeed become the past. There will no longer be anything like a *holocaust* in anyone's memory. Those that would so believe would be considered insane.

Religious groups and brainwashing techniques use the same principle. Get enough of a group to believe something, and it becomes the reality.

In the movie and the book *1984,* George Orwell presents us with the power of the quantum world view. *He who controls the past, controls the present. He who controls the present, con-*

trols the future. Thus by manipulating data from past records, a new past is created that eventually becomes the mainstream thought of Orwell's society. My point is that the past is a very shaky affair at best, and there may be no real past, only a past that we come to agree was the real past.

So it seems, although it appears preposterous, that we can create a past provided enough minds come to agreement. That which we call the past is that which most agree took place. I'm sure that the past as "remembered" by the followers of the old Shi'ite faith and its leader, Ayatollah Khomeini, is very different from the past remembered by the citizens of Iraq. Israelis have a different record of the past than Palestinian refugees. I don't just mean the personal memories of citizens. I mean the collective, agreed-upon memories of countries. I'm sure that within our own borders there are differing memories, different pasts for ethnic groups.

With the coming of quantum computer memories that specifically can alter the past state by creating a past state from merging parallel processing streams, the question becomes one of physical record as well. In a *Star Trek* episode, Captain Kirk is accused of negligence bordering on manslaughter; a computer record of his recent *yellow alert* aboard the *Enterprise* resulted in his court-martial. The accuser in this case was the computer, which showed clearly that Kirk pushed a button ejecting a pod containing a fellow officer. Such an ejection should have been done only in the case of a *red alert*. Kirk claims that the ejection did not take place during the yellow alert, and that a warning was issued to the officer to abandon the pod. The pod was then later ejected when the red alert was signaled and the officer was apparently safely out of the pod.

But the computer record clearly differs from Kirk's claims. In the future time of *Star Trek,* the computer is no doubt a quantum computer. It contains records that are not possible today, such as a complete holographic visual and sound record. The user can expand any single visual frame for more detail or use a scanning device to look at different parts of the ship, much as television cameras scan a scene. The control for the scan lies in the present, allowing records to be examined in great detail. Kirk is subjected to this scrutiny. The visual scan shows him pushing the eject button while a yellow alert is in progress.

But Kirk is freed and found innocent of wrongdoing when the defense attorney makes everyone realize that the computer is after all only a machine and not a human being. He claims that the records had to have been altered, which turned out to be true, by the officer in the pod to make Kirk look bad. The officer was actually still alive and hiding out.

What interests us here is the fact that the past was reconstructed by the hiding officer using quantum physical techniques that would indeed be able to make such a record appear as fact.

If quantum computers do make their way into our lives, then such a situation as past reconstruction will be a reality (indeed to some extent it already is a reality just using the present state of the art of computers). Perhaps the imagined *1984* will take place in the future when the calendar points to 2010. I surely hope not.

Chapter 31

TALKING TO TOMORROW'S PARALLEL UNIVERSE

If the reader has been bearing with me as we moved through this book, I am sure that many questions have arisen. By now I am sure that you see that physics itself is the attempt to logically figure out the illogical—perhaps an impossible task. What began as a theoretical means—a mathematical picture of reality in the days of Galileo and Newton—has turned into a modern abstractionist view of reality where nothing familiar exists on the canvas. Some feel that we physicists are holding back and that we really know what reality looks like. I can assure the reader that none of us has any hidden agendas for reality or any hidden views. We are trying to build a picture of the universe that fits self-consistently. Ultimately we believe that the universe is logical—but the logic may transcend our ordinary use of that word.

Indeed the new experiences we physicists are having as we

probe into deeper and more abstract questions require us to change the meanings of words and invent new ones all of the time. We need some new labels for reality.

In these final chapters I will attempt to define reality and existence, two words that I don't believe are synonymous. I shall also attempt to recap my viewpoint taken here: The parallel universes predicted to exist in the theory of general relativity are the same as those predicted in quantum physics. Following this line of reasoning we are drawn to the following conclusions:

1. There are an infinite number of parallel universes.
2. Quantum waves carry information moving from past to present and from future to present.
3. We should be able to "talk" to the future as clearly as we "talk" to the past.
4. Existence as we know it is a subset of reality which is unknowable.

The Infinite Number of Parallel Universes . . . Inside Your Head

Now, like the proverbial potato chip in the television ad, "You can't eat just one," you also can't have just one parallel universe. If there is one, there must be an infinite number of them. Thus in this logic, parallel universes imply quantum physics, not the other way around. Once we arrive at the quantum world view and accept its rather weird parallel universes prediction, we find that the whole picture appears self-consistent. The whole universe is connected through the existence of all these parallel universes.

Each and every atom in your brain and body is connected to each and every atom everywhere. Indeed the most profound connection is between the atom and its sisters or brothers in the local parallel universes that immediately surround it. Those universes make up its electron wave cloud. As long as no attempt is made to discern where the electron actually is, it is present in all of these universes simultaneously. These universes overlap, producing the single universe containing a stable atom.

When an electron is discovered in any universe by an observing piece of equipment or an atomic scientist, the overlapping universes split apart. Then that equipment or scientist enters into all the parallel universes, each sister piece or scientist's brain recording a different measurement for the electron's location. All of these measurements are consistent with the distribution predicted by the quantum wave function, which indicates how probable the electron is to be found at any one point in the atom.

The Scientist's Brain Is Split by the Atom

The fact that the scientist's brain has been split by the atom is usually not noticed. The whole ensemble of parallel brains occupies the same space in just one universe, and the split is virtually unnoticed. The rest of the universe simply doesn't care to note that this particular scientist has made that particular discovery. As far as the world is concerned, the entity they call "the scientist" or "the equipment" is just one entity, even though this entity has been split into an infinite number of parallel entities.

This is exactly the same as the situation for an unobserved atom. The electron in it exists in an infinite number of parallel universes. Only since no one has observed any of the electron's positions, the universes overlap. No one has attempted to measure in which universe the atomic electron resides. Thus all the atomic electronic universes that make up the wave cloud reside in this single universe, because no one cares to take notice.

The point is that any one universe is composed of virtually an infinite number of potential or unobserved overlapping parallel universes so long as no resolution or observation is attempted. In this manner the whole range of unobserved phenomena represents a completely undivided universe. This whole undivided mess is greater than the sum of its possibilities, because the various possibilities are able to interfere with each other at a level existing before observation occurs. It was this interference that made quantum physics so unique.

Since we believe in quantum physics, we are forced to consider its major conclusions very carefully, even though they run quite counter to our sense of security or well-being. One of these major conclusions is continuing to be investigated even as I write these words. That is the nature of time in quantum physics. It appears that time and mind are connected, and a good possibility for the arising of order and evolution in the universe comes from the future. Thus—

Information Flows from Past to Present and from Future to Present

What do I mean by information moving from past to present and from future to present? I refer to a picture we have of time existing as a river. We thus flow along this river always in the direction from past to future. Actually we could just as well think of the river flowing from future to past, and we sit in the boat holding our position against the current of time. As we sit and gaze at the view, various flotsam and jetsam in the river pass by us. Perhaps we see a bottle float by. We reach down in the river and pull up the bottle and find a message inside: "Greetings from the twenty-first century."

The best picture, according to quantum physics, is that information flows in both directions simultaneously. The river of time has two counterstreaming currents. Information coming from the future as well as from the past influences the present. Thus we pick up two bottles every time we reach down into that river, not one. And we find inside two messages.

However, these bottles are strange. They don't really exist until we reach down into the river. If we could "see" what the river actually contained we would see countless ghostly bottles ever streaming from the source of the river in the mountains of information piled up in our pasts and from the seas of future information telling us what the weather is like ahead. These counterstreaming bottles only become real in our hands. When we reach down into that river a past bottle and a future bottle coalesce into one bottle, and like a magical little genie in the bottle, a message instantly appears.

That message explains the situation at the present moment. It also contains an orientation map telling or mapping out what was the past and what will be the future.

Alas, the message cannot be taken to be the absolute truth. It is only a probable truth. The probability concerns both the past and the future. Only the experienced present moment is for sure.

As strange as this picture is, it is probably closer to the truth than we might imagine.

Chapter 32

ALPHA AND OMEGA

Despite its enormous applications, quantum physics is still very contrary to human experience. Einsteinian relativity with its predictions of black holes and the like is nearly as bad. These ideas just run contrary to our everyday experiences and thus are extremely hard to believe.

The main problem in quantum physics is its interpretation. How can we believe that there exists an infinite number of universes, one appearing every time anyone happens to observe something? And further, that these universes are not just possibilities, but possibilities that somehow conspire to produce the world we do experience?

And even if that idea is difficult to grasp, just think about the alternative. There is just one universe; however, in it there is still the problem of the observer effect—namely, that whenever an intelligent being observes some aspect of the physical world, that aspect, originally just one possibility among many, pops into existence.

Consequently, even though quantum physics is used every day in physics labs around the world, no one really understands it—particularly anyone who believes in an objective, causal, logical universe.

I believe that in spite of its bizarre appearance, the parallel universe theory offers the most satisfactory understanding of reality. My idea that the parallel worlds of quantum physics are the same as those that arose in the big bang and the same as those predicted to arise by the general theory of relativity also sheds some light on the possibility of detecting them. To detect a parallel universe we will need to face some new choices —to make some new decisions based on new knowledge.

Each day of our lives we make decisions. These decisions are sequences of events—histories of happenings, one following the other. In order to do so, we must proceed through life as if we had complete information. But the uncertainty principle forbids us from ever obtaining complete knowledge about any sequence of events. So it is somewhat of a paradox that we are able to proceed as if we had such complete knowledge.

The concept that these events take place willy-nilly with no one able to control or determine the exact result of any outcome is unsatisfactory. Using only the normal or Copenhagen interpretation of quantum physics forces us to regard the future as having no consequence for the present. This, going hand in hand with our classical physics understanding of the universe, puts the past on a special footing. It is only the past that rules the present, albeit somewhat poorly. The past in classical physics determines the present. Nowhere is there any room for a deciding mind.

The past in the Copenhagen interpretation of quantum physics plays a similar role. Only instead of determining the present, it sets the initial conditions for the quantum wave function; in other words, it sets limits on what could develop in the future not what will develop. That and nothing more.

This function, carrying information in the guise of probabilities, then propagates into the future following the quantum rules of wave propagation. Only when an observation occurs does this wave stop propagating and then suddenly collapse, when an event is recorded in memory, be it the memory of a recording device or of a human mind.

Thus there is no basis at all for the most common of all experiences—the ability to choose. We seem to be governed by random laws that mercilessly force us to regard the past as an inept guide to the future. The present occurs without our consent and seemingly beyond our control in this view of quantum physics. How do we get out of this seeming paradox? One way is to assume that both the past and the future exist simultaneously. But some physicists object to any conclusions one may draw about parallel universes.

Both the Future and the Past Exist

The view I have put forward in this book says that both the future and the past are real. The future plays a role in determining for us a basis upon which we may make decisions. But to do so, parallel paths to that future must also be real. If they are, the world is far more complex than we ever have imagined. God's work far exceeds any that human consciousness can do. Perhaps quantum physics is God's way of running the universe.

However, physicist Toyoki Koga, from Redondo Beach, California, in a debate about the existence of parallel worlds, took issue with the supposed sanctity of quantum theory.[16] He disagreed with the notion that quantum theory was all there was, and therefore thought that the many-worlds ideas were taken too literally. He believed that it is beyond our experimental means to distinguish one theory from another when dealing with such subjects as atoms and molecules. These elements are not directly observable, and so any theory about them cannot be relied on because it deals with things that are not directly perceived.

Another physicist, Joseph Gerver from Berkeley, California, in the same debate,[17] saw another paradox in the many-worlds theory due to the fact that the Everett theory makes no provision for branchings taking place moving backward toward the past as it does moving forward toward the future. Let me try to clarify Gerver's objection.

Parallel worlds should not only split as time goes forward whenever any observation or interaction between subatomic objects occurs, but, logically speaking, also evolve into splittings as time goes backward. Now by time running backward, he meant the same as watching a movie of the universe with its projector running backward. As he pointed out:

> *If it is possible for the universe to split into two slightly different realities by a quantum-mechanical event, then surely it is equally possible for two slightly different universes to become identical in the same manner.*

In other words, since we should see as many splittings occurring when we reverse time, then when time ran forward Gerver wants to see parallel worlds merging with the same frequency as the theory predicts that the words split apart. There is nothing in Everett's theory that would provide this. Then, if mergers take place with the same frequency as branches, by watching a movie of all possible branches of the universe running backward in time, one should see worlds branching off that look just like worlds running forward in time. Because one can keep following branches both backward and forward in time indefinitely, there must be an indefinite number of universes. This would imply that there is nothing special about this universe, and this bothered Gerver.

Universes branching off when watching the movie going

backward should appear the same as watching movies of the universes running forward. No doubt, some of the branches we would see looking backward will look like some of the branches we remember as the past. But the overwhelming majority of these past branches will appear like backward movies of the future. (In fact, a few of the "atypical" future branches will look like backward movies of the past.) Gerver pointed out, however, that the parallel worlds theory carries a saving grace. As he says:

> *If we accept multiple universes then we no longer need worry about what* really *happened in the past, because every possible past is equally real. Therefore, to avoid... insanity, we can, with clear consciences, arbitrarily define* reality *as that branch of the past that agrees with our memories.*

The Omega Point

Physicist Bryce DeWitt considered both Koga's and Gerver's comments. To Koga he replied that nothing sacred was implied by the parallel world idea. It is just the best theory we have come up with to date. To Gerver, he replied that Gerver's whole argument was based on the existence of a special point in time—an *omega point*—where time exhibited a mirror symmetry. The parallel worlds theory does not need such a point. He doesn't agree with the Gerver view that the past history of the universes (not to be confused with our own past history, which involves only one branch) would look like "backward movies of the future." This would be the case only if the present universe was the result of a fluctuation from a state of equilibrium in an infinitely old universe—in other words, our universe were nothing more than a fluke of time.

There are many reasons for believing that this is not the case. The big bang scenario imposes a directionality to time.

Thus the time development of the many universes should not necessarily have a point of time symmetry.

But suppose there was such a point in time. The big bang could be such a point. This would mean that just before the big bang, there was a big crunch. But for those living through the crunch time just before the end of their universe as they knew it (they would be crushed out of existence and wouldn't know that relief was just a bang away), the experience would be time-reversed. They would experience the big crunch as a big bang, too. In fact, assuming that the big bang were a time symmetry point, everything leading to the crunch would be an exact time-reversed image of everything happening after the bang. The two universes would be so identical that no one could tell them apart.

If the time symmetry point is placed at the moment of maximum expansion, there will be worlds (branches) in which time flows one way and worlds in which time flows the other way. There will also be a few maverick worlds where time might even oscillate, first flowing one way and then returning to the same conditions it flowed from.

All of these worlds would be unaware of one another unless there were quantum interference between them, as in the case of the double-slit experiment and the quantum coin of sides and colors. In each case the number of branches into which a given world splits increases in time but not if time runs backward. In fact, the direction of increasing splits would be taken as the direction of time. Thus if there were a point of "great symmetry," we would never know it.

An Overall View

Taking an overall view of this complexity leads to regarding the quantum wave function as a real wave. And if it is real, it exists in all of the parallel worlds simultaneously. It never stops propagating. Indeed it appears that it propagates both from the past (that impossible-to-imagine past we have come to call the "big bang" or "alpha point") and from the future (sometimes referred to as the "omega point").

From alpha to omega we have the one universal wave propagating through endless pathways—histories of events recorded in any form of consciousness that happens to be around. We also have the quantum wave propagating from the omega to the alpha, going against the time stream carrying information about the future all the way back to the past alpha. These two waves continually clash against each other. They are nearly always in opposition to each other.

But from time to time across the immense range of time, starting with alpha and ending with omega, these waves happen to coincide—they meet in phase with each other. Coincident waves tend to bundle into groups. These groups become parallel universes. Each parallel universe is much like the other, but there are differences depending on the amount of coincidence. The difference between one universe and the other is measured by the lack of coinciding waves.

In a universe, the places where waves coincide, events occur. Something physical appears. And with the appearance of a physical object, a subatomic particle, an atom, a molecule, a group of molecules, a structural arrangement of molecules, a cell, a group of cells, and practically everything else in what we call the universe, there is also something else. That something else is intelligence, order, and consciousness.

There is associated with every appearance of a physical object a site of knowledge. When these sites of knowledge are grouped together in a tight region of space alone, we have consciousness.

Thus consciousness is the resonant gathering of simultaneously present quantum waves—clashing waves of time—one coming from the future and one coming from the past. This defines what we mean by the present or now. And in this manner mind emerges as the focal point of waves coming from both the future and the past.

Looking at the human mind for a moment, we can use this picture to construct how it is that we know something.

Knowledge of a single event, isolated from all other events, is no knowledge at all. Knowledge without a sense of the past and an anticipation of the future is no knowledge at all. In brief, single data points without any reference to the future or the past are not even recorded as data. They simply do not exist. Only when there is confluence of past and future—a relationship stretching across time—does knowledge present itself.

Knowledge or mind is then the relationship of future to past—that relationship appearing as, and creative of, the present moment. Knowledge is thus equivalent to meaning, and meaning is only present when there are two events—a future and a past. Knowledge, meaning, and time order then become synonyms. To know *is* to put events in a time order *is* to have meaning. To have meaning *is* to relate a pair of events. That pair is always oriented so that one event is the past and the other is the future. The place where this knowledge occurs is the present. That which recognizes the present is mind.

Mind emerges as the organ that best recognizes the past and the future. A mind without a past is no mind at all. A mind without a future is no mind at all. Mind cannot exist with both past and future occurring simultaneously. It is in the time separation that mind arises. This is the meaning of the present moment.

From the point of view taken here, that of quantum physics, the past alone cannot define the present because the present cannot be predicted from the past alone. A past determination of an event is no security. A measurement of the position of an object in the past leads to no knowledge of the momentum of the object, and therefore no knowledge of the presence of the object.

What must be added to the picture is the future. Yet, the future measurement of the momentum alone is no basis for the retrodiction ("prediction" of the past) of the position of the object, and again no basis for the knowledge of the presence of the object. But taken together, both the past position and the future momentum coincide in the present and the object "now" exists. The future momentum and the past position together create a simultaneously knowable presence of both position and momentum.

And this coinciding of complementary attributes is required in order that our minds gain predictable power about anything that physically appears. The predictability of anything depends on the simultaneous knowability of complementary observables. Perhaps our abilities to make sense of the present depend not only on what we have accomplished in the past, but also what we are committed to accomplish in the future.

Take away the past or the future and our lives become meaningless.

The world we see out there appears in physical form because information from the past and from the future, containing complementary observations of the object in question, joins for a momentary flash of consciousness. Although the present moment can only be partially known through the observation of either one of the complementary observables, nevertheless, both attributes are "present" and knowable. This is why the world appears so ordinary and stuffed chockablock with material. Throw out either the alpha or the omega, and consciousness also vanishes—the world would become ghostlike and nothing would exist as a solid object.

Chapter 33

REALITY AND EXISTENCE

So what is reality? What is existence? From my quantum viewpoint, the viewpoint taken in this book, the universe, any universe, all of the parallel universes are real, but not all exist. Here in the finale I wish to separate reality from existence and describe how parallel universes fit into the discussion.

First of all, let us consider what we mean by reality. *The American Heritage Dictionary of the English Language* describes reality as:

1. The quality or state of being actual or true.
2. A person, entity, or event that is actual.
3. The totality of all things possessing actuality, existence, or essence.
4. That which exists objectively and in fact.
5. The sum of all that is real, absolute, and unchangeable.

The definition of reality that I shall use here adopts some of the above but not all. For me, reality consists of a gigantic superspace—the mathematical space of all possibilities. We might think of this as the *mind of God.* Each dimension in the space is like a dimension in our ordinary three-dimensional space. Each dimension of the space also points or indicates a "direction." This direction is a possibility. In this space something called mind floats and freely associates. It puts together—for a fleeting moment, which could be an eternity, depending on the time scale one is looking at—any two or more dimensions, and it asks what about this combination?[18]

Our minds are thus tuned or are tunable to multiple dimensions, multiple realities. The freely associating mind is able to pass across time barriers, sensing the future and reappraising the past. Our minds are time machines, able to sense the flow of possibility waves from both the past and the future. In my view there cannot be anything like existence without this higher form of quantum reality.

I believe that this insight into the workings of quantum physics, a view that I have taken based on the work of several other physicists, including John Cramer, John A. Wheeler, Sir Fred Hoyle, David Z. Albert, Yakir Aharonov, Susan D'Amato, Jack Sarfatti, and many others, is the most important insight into this strange landscape that has occurred since the discovery of quantum physics in the first place.

If it turns out to be a testable hypothesis, it will revolutionize our view of the world. It will say in effect that time is not a barrier. The future exists now, and so does the past.

Finally, from this perspective much is evident. We are both the choosers of subsets of reality and the constructors of existence. Because we are the products of time past and time future we are "stuck" in time.

This stickiness is God's way of having anything real become something in existence. It must stick into the double flow of transactions as the ultimate middleman receiving profits from both ends simultaneously. The more resonant the transaction, the higher its probability and the greater the number of parallel universes contributing to the transactional existence. Since these universes are constantly overlapping and splitting through time, we are ultimately the beneficiaries or the "maleficiaries" (receivers of bad gifts) in this parallel universe game.

As T. S. Eliot put it in *Four Quartets:*

> *Time present and time past*
> *are both perhaps present in time future,*
> *and time future contained in time past.*
> *If time is eternally present*
> *All time is unredeemable.*
> *What might have been is an abstraction*
> *Remaining a perpetual possibility.*
> *Only in a world of speculation,*
> *What might have been and what has been*
> *Point to one end, which is always present.*
> *The end of all our exploring*
> *will be to arrive where we started*
> *And know the place for the first time.*

Indeed the action of piercing through the time fog and the existence of parallel universes appear to go hand in hand. The fog of time is parallel universes. One of my views is that it will be possible to have experiences of the future. We already know of the existence of "psychics" and seers, and many of us scoff at the idea while others have no difficulty in accepting them.

So if time is present and unredeemable, might it be possible to observe a future parallel universe?

The only way to have time unredeemable is to have it exist as a maze of parallel universes. Perhaps a better image is the hologram of parallel universes.

What one sees when viewing the hologram depends on the viewpoint one takes. And that viewpoint alters the hologram because it alters the probabilities by moving in time.

After all, if the hologram is constructed from transacting quantum waves moving in all time directions, then who assigns the probabilities? Somehow a field of consciousness must illuminate the hologram and *IT* assigns the probabilities.

There must be a conscious observer. In my view this observer is ourselves—a spill from the giant ocean of thought that is God, temporarily trapped by the hologram, but also unredeemable and unchanging.

Woody Allen's comment, in the introduction of the book, about unseen worlds reminds us that midtown is between the poles of the future and the past. It also reminds us that we can go nowhere with such ideas unless there are directly observable consequences. We must be able to know how late it is open and whether a good corned-beef sandwich is to be found.

I believe that there are (both good sandwiches and observable consequences). The human mind is the laboratory of the new physics. It already is tuned to the past and the future, making existential certainties out of probable realities. It does this by simply observing. Observing oneself in a dream. Observing oneself in this world when awake. Observing the action of observing. If we are brave enough to venture into this world with consciousness as our ally, through our dreams and altered states of awareness, we may be able to alter the hologram by bringing more conscious "light" to the hell worlds that also exist side-by-side with our own.

Indeed what with our bent toward constantly "defending ourselves," this parallel world already is a hell world. It is time to speed up the process of illuminating the hologram, time to bring in the big laser of consciousness. Evolution is our business too. It is time to know this universe-place for the first time ever.

NOTES

Part One

1. This story is found in
   Bradbury, Ray. *The Silver Locusts: A classical collection of S.F. stories taken from the Martian Chronicles.* London: Corgi Books, 1956.
2. See the exposition of Everett's ideas in
   DeWitt, Bryce S., and Graham, Neill. *The Many-Worlds Interpretation of Quantum Mechanics.* Princeton, New Jersey: Princeton Univ. Press, 1973.
3. This story is found in
   Borges, Jorges. *Ficciones.* New York: Grove Press, 1962, p. 100.
4. New Techniques and Ideas in Quantum Measurement Theory, conference held by the New York Academy of Sciences on January 21–24, 1986, in New York City.
5. Thomsen writes well about quantum physics for *Science News.* A number of his articles about this subject are listed below.
   Thomsen, Dietrick E. "A Knowing Universe Seeking to be Known." *Science News.* Vol. 123 (Feb. 19, 1983), p. 124.
   ______. "The Quantum Universe: A Zero-Point Fluctuation?" *op. cit.* Vol. 128 (1985), p. 72.
   ______. "Going Bohr's Way in Physics." *op. cit.* Vol. 129 (1986), p. 26.

———. "Holism and Particlism in Physics." *op. cit.* Vol. 129 (1986), p. 70.
———. "Quanta at Large: 101 Things to Do with Schrödinger's Cat." *op. cit.* Vol. 129 (1986), p. 87.
———. "Notes of an Ex-Physics Student." *op. cit.* Vol. 129 (1986), p. 141.

6. My earlier book also discusses Bohr's Copenhagen interpretation. See Wolf, Fred Alan. *Taking the Quantum Leap: The New Physics for Nonscientists.* San Francisco: Harper & Row, 1981.
7. See
   Bohm, D. J., Dewdney, C., and Hiley, B. H. "A Quantum Potential Approach to the Wheeler Delayed Choice Experiment." *Nature.* Vol. 315 (1985), p. 294.
   Dewdney, C., Gueret, Ph., Kyprianidis, A., and Vigier, J. P. "Testing Wave-Particle Dualism with Time-Dependent Neutron Interferometry." *Physics Letters.* Vol. 102A, No. 7 (1984), p. 291.

Part Two

1. A word about such notation as 10^{11}. It means the number 10 multiplied by itself 11 times. In words that's one hundred billion.
2. See
   Barrow, John D., and Tipler, Frank J. The *Anthropic Cosmological Principle.* New York: Oxford University Press, 1986.
3. See
   Ballentine, L. E., Pearle, P., Walker, E. H., Sachs, M., Koga, T., Gerver, J., and DeWitt, B. "Quantum Mechanics Debate." *Physics Today.* April 1971, p. 36.
4. See
   DeWitt, Bryce S. "Quantum Mechanics and Reality." *Physics Today.* Sept. 1970, pp. 30–35.
5. See note 3.
6. See note 3.
7. See note 3.
8. See note 4.
9. The Dirac equation is a mathematical expression developed by physicist Paul Dirac in order to explain the behavior of electrons moving very near to the speed of light. Dirac discovered that all particles of matter move at the speed of light following jagged paths through space. This "jitterbugging" motion produces the illusion that matter is moving slower than light. He also showed that every subatomic particle is capable of existing below the threshold of any perception and that an infinite number of like particles must exist at that level. When certain energies are created, one of these particles can be made to manifest out of nothing, leaving behind a hole. This hole also has physical properties and appears as the antiparticle of the particle that manifests.

Part Three

1. Einstein, A., and Rosen, N. "Particle Problem in the General Theory of Relativity." *Physical Review.* Vol. 48 (1935), pp. 73–77.
2. Consider the possibility that our universe is a black hole. To come to this conclusion we need to consider the factors that define a black hole. They are mass, M, radius, R, and density, D. The radius of a black hole is directly proportional to its mass ($R \sim M$). However the density of a "thing" like a black hole is given by its mass divided by its volume ($D = M/V$). Since the volume of a black hole is proportional to the radius of the black hole to the power three ($V \sim R^3$), the density of a black hole is inversely proportional to its mass raised to the second power ($D \sim M^{-2}$). That means if a black hole has a lot of mass it doesn't appear very dense!

   Our universe is certainly not very dense, being made up mostly of space. But there is quite a bit of mass in it. Thus it would appear consistent that our universe being quite massive but not dense could indeed be a black hole.
3. In mathematics, we find a certain class of numbers called imaginary numbers. Before we look at them we need to look at numbers that are related to them. These are called *square roots.* Now a square root isn't too difficult to grasp. For example, the square root of 4 is 2. The square root of 9 is three. Thus a square root of a number (e.g., 4) is another number (e.g., 2) that you multiply by itself (2×2) to get the original number (4). The square root of 1 is, however, also 1. One is strange in this regard. Square it and you get 1. So taking the square root also gives 1. (This just means $1 \times 1 = 1$, or $1^2 = 1$.)

   But does any number have a square root? In other words, does there always exist a number that when multiplied by itself gives another number? You would certainly think so. How about negative numbers? Negative numbers appear when you overdraw your bank account. Can overdraws have square roots, too? What does taking the square root of minus one mean? It means that you get a number that when multiplied by itself gives minus one. Now we don't have such numbers in the real world. For example, I don't have a square root of minus one dollar in my pocket. I could have the square root of four dollars in my pocket (two bucks). But a minus one dollar? Nevertheless, it is meaningful to use such a number because when you square it you do get a number that is useful. I could have a minus one dollar in my pocket. It would mean I had an IOU of one and owed someone a dollar.

   Whenever a number is useful but we don't have a symbol for it, we invent one. In the case of the square root of minus one, the symbol is i. This just means that $i \times i = -1$. We call all numbers multiplied by i imaginary numbers. Thus $i5$ is imaginary five and $(i5)^2 = -25$.

Minkowski noticed that if you insert *i* as a multiplier of the time symbol in Einstein's equations, and you measure all speeds in relationship to the speed of light, you can reproduce all of Einstein's relationships and give them a visual, geometrical sense. In doing this the time that we knew became an imaginary dimension of space. Using imaginary space and real space, all of the Einstein relationships could be drawn in the form of triangles. All one had to do was let the time leg of the triangle be an imaginary space dimension and the space leg of the triangle be a real space dimension.

4. Clark, Ronald W. *Einstein: The Life and Times.* New York: World Publishing Co., 1971, p. 159.
5. See note 1.
6. Kerr, Roy P. "Gravitational Field of a Spinning Mass as an Example of Algebraically Special Metrics. *Physical Review Letters.* Vol. 11 (1983), pp. 237–38.

Part Four

1. Weinberg, Stephen. *The First Three Minutes.* New York: Basic Books, 1977.
2. See
   Wolf, Fred Alan. *Taking the Quantum Leap.* San Francisco: Harper & Row, 1981.
   ———. *Star Wave.* New York: Macmillan, 1984.
   ———. *The Body Quantum.* New York: Macmillan, 1986.
3. Reported by Dietrick E. Thomsen in the May 30, 1987, issue of *Science News,* Vol. 131, p. 346.

Part Five

1. See
   Wheeler, John A. "Delayed-Choice Experiments and the Bohr-Einstein Dialogue." Paper read at a joint Meeting of The American Philosophical Society and the Royal Society, June 5, 1980, London. Library of Congress catalog card number 80-70995, 1980.
   Tipler, Frank J. "Rotating Cylinders and the Possibility of Global Causality Violation." *Physical Review D.* Vol. 9 (1974), p. 2203.
   ———. "Interpreting the Wave Function of the Universe." *Physics Reports.* Vol. 137 (May 1986), p. 4.
2. See
   Vonnegut, Kurt. *Slaughterhouse Five.* New York: Dell Publishing Co., 1969.
3. Cramer, John G. "Generalized Absorber Theory and the Einstein-Podolsky-Rosen Paradox." *Physical Review D.* Volume 22 (1980), p. 362.
   Also see

———. "The Transactional Interpretation of Quantum Mechanics." *Reviews of Modern Physics*. Vol. 58, No. 3 (July 1986).

4. See

Hellmuth, T., Zajonc, Arthur C., and Walther, H. "Realizations of Delayed Choice Experiments." In *New Techniques and Ideas in Quantum Measurement Theory,* ed. D.M. Greenberger, Vol. 480, Annals of the New York Academy of Sciences, December 30, 1986.

Wheeler, John A., "The Mystery and the Message of the Quantum." Presentation at the Joint Annual Meeting of the American Physical Society and the American Association of Physics Teachers, Jan. 1984.

———. In *The Mathematical Foundations of Quantum Mechanics,* edited by A. R. Marlow. New York: Academic Press, 1978.

———. "Beyond the Black Hole." In *Some Strangeness in the Proportion: A Centennial Symposium to Celebrate the Achievements of Albert Einstein.* Edited by Harry Woolf. Reading, Mass.: Addison-Wesley, 1980, p. 341.

———. "Delayed-Choice Experiments and the Bohr-Einstein Dialogue." Paper read at a joint Meeting of The American Philosophical Society and the Royal Society, June 5, 1980, London. Library of Congress Catalog card number 80-70995,1980.

5. See

Allman, William F. "Newswatch: The Photon's Split Personality." *Science 86*. June 1986, p. 4, for a popular description of this experiment.

Also see

Hellmuth, T., Zajonc, Arthur C., and Walther, H. "Realizations of Delayed Choice Experiments." In *New Techniques and Ideas in Quantum Measurement Theory,* ed. D.M. Greenberger, Vol. 480, Annals of the New York Academy of Sciences, December 30, 1986.

6. For a description of the double-slit experiment see Chapter 2 in Part One.

7. See

Tipler, Frank J. "Rotating Cylinders and the Possibility of Global Causality Violation." *Physical Review D*. Vol. 9 (1974), p. 2203.

Part Six

1. Le Guin, Ursula. *The Lathe of Heaven.* New York: Avon. 1973.

2. See the article by Leon Cooper and Deborah Van Vechten in

DeWitt, Bryce S., and Graham, Neill. *The Many Worlds Interpretation of Quantum Mechanics.* Princeton, New Jersey: Princeton Univ. Press, 1973.

3. See

Hoyle, Fred. *The Intelligent Universe.* New York: Holt, Rinehart & Winston, 1983.

———. "The Universe: Past and Present Reflections," *Preprint Series No.*

70 Astrophysics and Relativity. Cardiff, United Kingdom: Department of Applied Mathematics and Astronomy, University College, May 1981.

Also see

Aharonov, Yakir, and Albert, David Z. "Is the Usual Notion of Time Evolution Adequate for Quantum-Mechanical Systems? I." *Physical Review D.* Vol. 29 (1984), p. 223.

Aharonov, Yakir, Albert, David Z., and D'Amato, Susan S. "Multiple-Time Properties of Quantum-Mechanical Systems." *Physical Review D.* Vol. 32 (1985), p. 32.

Albert, David Z., Aharonov, Yakir, and D'Amato, Susan S. "Curious New Statistical Prediction of Quantum Mechanics." *Physical Review Letters.* Vol. 54 (1985), p. 5.

Cramer, John G. "Generalized Absorber Theory and the Einstein-Podolsky-Rosen Paradox." *Physical Review D.* Vol. 22 (1980), p. 362.

———. "The Transactional Interpretation of Quantum Mechanics." *Reviews of Modern Physics.* Vol. 58, No. 3 (July, 1986).

4. See

Aharonov, Yakir, Albert, David Z., and D'Amato, Susan S. "Multiple-Time Properties of Quantum-Mechanical Systems." *Physical Review D.* Vol. 32 (1985), p. 32.

D'Amato, Susan S. "Two-Time States of a Spin-Half Particle in a Uniform Magnetic Field." In *New Techniques and Ideas in Quantum Measurement Theory,* ed. D.M. Greenberger, Vol. 480, Annals of the New York Academy of Sciences, December 30, 1986.

5. See

Albert, David Z. "On Quantum-Mechanical Automata." *Physics Letters.* Vol. 98A (1983), pp. 249–252.

Deutsch, D. "Quantum Theory, the Church-Turing Principle and the Universal Quantum Computer." *Proceedings of the Royal Society of London,* Vol. A 400 (1985), pp. 97–117.

6. New Techniques and Ideas in Quantum Measurement Theory, conference held by the New York Academy of Sciences on January 21–24, 1986, in New York City.

7. Hoyle, Fred. *October the First Is Too Late.* London: Wm. Heinemann, 1966.

8. See

Hoyle, Fred. "The Universe: Past and Present Reflections," *Preprint Series No. 70, Astrophysics and Relativity.* Cardiff, United Kingdom: Department of Applied Mathematics and Astronomy, University College, May 1981.

9. See

Aharonov, Yakir, Albert, David Z., and D'Amato, Susan S. "Multiple-Time Properties of Quantum-Mechanical Systems." *Physical Review D.* Vol. 32 (1985), p. 32.

Albert David Z., Aharonov, Yakir, and D'Amato, Susan S. "Curious

New Statistical Prediction of Quantum Mechanics." *Physical Review Letters.* Vol. 54 (1985), p. 5.

D'Amato Susan S. "Two-Time States of a Spin-Half Particle in a Uniform Magnetic Field." In *New Techniques and Ideas in Quantum Measurement Theory,* ed. D.M. Greenberger, Vol. 480, Annals of the New York Academy of Sciences, December 30, 1986.

10. See

    Albert, David Z. "How to Take a Photograph of Another Everett World." In *New Techniques and Ideas in Quantum Measurement Theory,* ed. D. M. Greenberger, Vol. 480, Annals of the New York Academy of Sciences, December 30, 1986.

11. See

    Deutsch, D. "Quantum Theory, the Church-Turing Principle and the Universal Quantum Computer." *Proceedings of the Royal Society of London.* Vol. A-400 (1985), pp. 97–117.

12. See

    Wolf, Fred Alan. *Star Wave: Mind, Consciousness, and Quantum Physics.* New York: Macmillan, 1984.

13. See

    Cooper, Leon N. and Van Vechten, Deborah. "On the Interpretation of Measurement Within the Quantum Theory." *American Journal of Physics.* Vol. 37 No. 12, (1969), p. 1212.

14. See

    Ballentine, L. E., Pearle, P., Walker, E. H., Sachs, M., Koga, T., Gerver, J., and DeWitt, B. "Quantum Mechanics Debate." *Physics Today.* April 1971, p. 36.

15. See note 13.
16. See

    Ballentine, L. E., Pearle, P., Walker, E. H., Sachs, M., Koga, T., Gerver, J., and DeWitt, B. "Quantum Mechanics Debate." *Physics Today.* April 1971. p. 36.

17. See note 15.
18. It does this in a manner similar to a schoolteacher holding a pointer. By orienting the pointer so that it points straight up, the teacher is looking at one dimensional possibility, up-down. By holding the pointer so that it points perpendicular to the north wall, she considers another dimensional possibility, north-south. By holding it so that it points along a diagonal, the teacher takes into account more than one of the prescribed dimensions, up-down and north-south, for example. By swinging the stick in the room, these possibilities are constantly changing, as quickly as the direction of the pointer changes.

    But the teacher is not alone. There is another teacher in a room nearly identical with that of the original teacher. The parallel teacher is somewhat of an automaton. He can only mimic or attempt to duplicate the movements of the original teacher. Let's call this parallel teacher the teacher-2. Each time the teacher points in a direction $|a\rangle$, the teacher-2 attempts to point in the same direction but usually misses the mark and points in a direction $\langle b|$. The direction $|a\rangle$ is called a *ket* and the direction $\langle b|$ is called a *bra*. Taken together the *bra* and *ket* make the word *bracket* and the association of *b* and *a*, $\langle b|a\rangle$. This is a fleeting association—a joining together of two ideas in the brain. This may be what takes place in the *id*. There free associations occur all of the time. Most of them are meaningless. This association is fleeting, but it is an element of reality. It is the association of possibilities *a* and *b*.

If the teacher and the teacher-2 can "get it together," and repeat the dance, then a pattern emerges where first the teacher points to *a*, the teacher-2 points to *b* and next the teacher points to *b* followed by the teacher-2 pointing to *a*. The dance goes, $\langle a|b\rangle\langle b|a\rangle$. Once this dance begins, the association of *a* and *b* becomes more than a reality, it comes into existence. This dance is beyond time. It is a double flow from a past *a* to a future *b* and then from a future *b* to a past *a*.

One more point, the teacher moves the pointer from the past to the present moment in time, while the teacher-2 moves his pointer from the future to the present. The double association of *a* and *b* is the probability of association of *a* and *b*. It can be viewed as a statement of logic, *If a then b,* when seen from the normal perspective of time running from past to present (the teacher point of view), or it can be viewed as the logic statement, *If b then a,* when viewed from the time-reversed perspective (the teacher-2). The important aspect of this is that a meaningful connection is made between two possibilities. If *a* and *b* are identical, the probability becomes a certainty and the possibility becomes an actuality. It comes into existence.

Now existence is defined by *The American Heritage Dictionary of the English Language:*

1. *The fact of having being or actuality.*
2. *The fact or state of continued being; life.*
3. *All that is present under certain circumstances or in a specified place.*
4. *A thing that exists, an entity.*
5. *A mode or manner of existing.*
6. *Occurrence or specific presence.*

In my view, existence is a result of a double flow of information. One stream comes from the past and the other comes from the future. A thing is said to exist in a state *a* given that it had existed in state *b*, by the product, $\langle a|b\rangle\langle b|a\rangle$. The product contains two factors multiplied together. It says both *If a then b,* and it says *If b then a.*

GLOSSARY

Anthropic Principle: States that from an infinite number of possibilities nature could have selected to make a universe, it selected this one so that we could be created.

Attosecond: A billionth of a billionth of a second. One attosecond compares with one second as one second compares with about thirty-two billion years.

Big Bang: A gigantic explosion at the beginning of time when all matter, energy, space, and time were suddenly created.

Black Hole: A spherical region of space which contains a gigantic gravitational field. The field is so large that everything on its surface is sucked into it, including light. Imagine a sphere that, like a magnet, attracts everything around it. Now let it suck in sunlight and you have a black hole.

Boundary Conditions: A special set of constraints or limitations on a physical system occurring naturally or artificially imposed by human means. Boundary conditions limit the movements of quantum waves by denying the quantum wave access to all of physical space. Accordingly, the quantum wave must be zero at such places where it's impossible to observe the physical quantity represented by the quantum wave. Altering boundary conditions usually leads to changes of values associated with physical measurements. Because the quantum wave influences probabilities, the mind may be capable of altering and changing matter by changing a quantum wave's boundary conditions.

CHRONON: The billionth part of the billionth part of the billionth part of the billionth part of the billionth part of one second. Now to find your way through all these billionths, remember that one billion seconds is just under thirty-two years. The big bang took place in the first chronon. But a billion chronons passed before there was any light in the universe. Thus in terms of chronons, something equivalent to thirty-two years passed before the first light emerged from the first point of space and time. One second is to 32 billion billion billion billion years as one chronon is to one second.

CLASSICAL MECHANICS: The laws of motion as conceived by Sir Isaac Newton. There are three laws:

1. A body in motion tends to stay in motion—the principle of inertia.
2. A force acting on a body will cause that body to accelerate—either speed up, slow down, or change its direction through space.
3. A force acting on a body will cause that body to return an equal and oppositely directed force on the source of the original force.

CLASSICAL PHYSICS: The laws of physics based on ideas that were in existence before quantum physics was discovered. These include classical mechanics, electricity and magnetism as theorized by James Clerk Maxwell, thermodynamics, and other branches of physics that are based on the aforementioned concepts. The laws of relativity are sometimes included in classical physics because they too are based on prequantum-physics concepts.

COLLAPSE OF THE WAVE FUNCTION: The sudden change in the quantum wave function when an observation takes place. Since the wave function represents the probability of observing an event, the collapse means that the probability has changed from less than certainty to certainty. *See* observer effect.

COMPLEMENTARITY: The principle that says the physical universe can never be known independently of the observer's choices of what to observe. These choices fall into two distinct or complementary sets of observations called observables. Observation of one observable always precludes the possibility of simultaneous observation of its complement. The observation of the location and the observation of the path that a moving subatomic particle is following are complementary observables.

COPENHAGEN INTERPRETATION: The interpretation first given by Niels Bohr, who is considered, in good quantum physical paradoxical language, the father and the mother of quantum physics. In the Bohr school, objects no longer have the same attributes as pictured by Newtonian physics. Objects possess two kinds of observables: those that can be observed simultaneously and those that cannot. Bohr's interpretation gave a reason for this behavior. He said that tiny objects aren't what large objects seem to be. A large object follows Newton's laws. Its path and location are simultaneously observable. But atom-sized objects are disturbed by any attempt to observe them. For example, if one carefully performs an experiment to observe an electron's position, that experiment necessarily blurs its path. Conversely, any experiment to determine the electron's momentum makes it impossible to determine its location. Bohr believed that this wasn't due to any ineptness on the part of the experimenter; rather, it was due to the inevitable consequence that eventually a large object, such as a machine, recording device, or human being, had to observe a tiny object, such as an electron or atom. The large object followed Newton's laws while the atom-sized object didn't. Since any information about the tiny object had to be obtained by a large object, all it could do was disturb the little particle, with unpredictable results.

COSMOLOGY: The theory of the early universe—how all that we can imagine as physical began about fifteen billion years ago.

DIMENSIONS:

REAL DIMENSION: A unique extension into space. While standing, stretch your right arm out in front of you and your left arm straight out from your side. Your right arm corresponds to the x-dimension, your left arm the y-dimension, and your body points in the z-direction. These are the three dimensions of space.

IMAGINARY DIMENSION: While standing there with your arms outstretched, experience the change around you that is the passing of time. This is movement through an imaginary dimension.

DIRAC'S EQUATION: A mathematical expression invented by physicist Paul Dirac in order to explain the behavior of electrons moving at very near the speed of light. Dirac discovered that all particles of matter actually moved at the speed of light following jagged paths through space. This "jitterbugging" motion produces the illusion that matter is moving slower than light. He also showed that every subatomic particle is capable of existing below the threshold of any perception and that an infinite number of like particles must exist at that level. When certain energies are created, one of these particles can be made to manifest out of nothing, leaving behind a hole. This hole also has physical properties and appears as the antiparticle of the particle that manifests.

DOUBLE-SLIT EXPERIMENT: The major interpretational experiment of quantum physics. In the experiment, a series of single particles, each emitted from a source one at a time, encounters a barrier containing two parallel slit openings. According to classical physical concepts, each particle must pass through one slit or the other if it is to reach a recording device on the other side of the barrier. However, the record of the spots made by one particle following another indicates that each particle must have passed in a wavelike manner through both slits at the same time—if no one checks which slit each particle passes through. On the other hand, if someone observes the particle passing through one slit or the other, the record of spots is altered. Thus the particle passes as a wave if no one looks, and it passes as a particle if someone sees.

Echo Wave: A quantum physical wave of probability that travels backward in time from the future carrying an imprint of the offer wave (*see* offer wave) seeking a present event to transact with (*see* transactional interpretation). The stronger the echo, the greater the likelihood both events will occur.

Einstein–Podolsky–Rosen Paradox: Deals with a measurement performed on one part of a physical system while the other part, which had previously been connected to it, is left alone. According to quantum rules, the measured part instantly affects the unmeasured part at the moment of measurement even though there is no longer any connection between the parts.

Electron: The smallest subatomic particle. The electron has certain measurable properties. These include electrical charge, inertial mass or resistance to accelerated motion, spin (which can be thought of roughly by picturing the electron as a tiny ball spinning about an axis), and electron exclusion (the tendency to avoid another electron by not entering the same quantum physical state) which appears whenever two or more electrons are near each other.

Equations of motion: *See* classical mechanics.

Event horizon: The surface of sphere marking the edge of what is known as a black hole. It is also called the sphere with the critical or Schwarzschild radius. It is called a horizon because like a horizon at sunset you can only approach it but never actually reach it. It takes an infinite amount of time to approach the event horizon, as measured by observers watching the spectacle from a distant viewing spot. However, only a finite amount of time is observed by the person approaching it. Again relativity of time shows its head. If you do manage to cross the event horizon, and enter into the interior of the black hole you can never turn back and reach the event horizon again. You will be swept away by the flow of spacetime inside the hole eventually ending your existence at the black hole's center.

Gas Dynamics: The laws that govern the physical properties of gases. These laws are usually based on Newtonian mechanics and thermodynamics.

Ground State: The lowest state of energy of a physical system as determined by the laws of quantum physics. This amount of energy is never zero. According to quantum physics, a physical system must always possess a residual amount of motional energy, even if it is cooled to absolute zero temperature.

Hologram: A device capable of forming an image of an object that appears to exist in full three dimensions. The image can be viewed from many angles, and with each change in viewing angle, the image of the object changes.

INFINITY: The concept that no matter how far you count, extend, or imagine, there is always one more.

INFLATIONARY PHASE: The theory that the universe, just after the moment of creation, expanded faster than light. This idea helps explain how the universe has come to be so consistent as observed via the background radiation detected as radio noise.

INTERACTION: A physicist's way of describing any two more objects that are mutually influencing each other. According to classical physics, both objects are, before and after the interaction, completely determined. All the interaction does is change the directions, positions, and momenta of the objects. According to quantum physics, before the interaction, either position or momentum of the objects is undetermined. After the interaction, the same is true for both objects, but they are, nevertheless, correlated. What one does to one of the objects instantly affects the other, even if the second object is not disturbed.

INTERFERENCE: The combining of two or more wave patterns superimposed one upon the other. *See* superposition. There are two types of interference.

DESTRUCTIVE INTERFERENCE: If the crests of one wave coincide with the valleys of another, the waves are said to destructively interfere with each other.

CONSTRUCTIVE INTERFERENCE: If the crests of one wave coincide with the crests of another, the waves are said to constructively interfere with each other.

INVARIANCE: A constancy that remains whenever something changes. Imagine moving through a wet rain forest and getting soaked to the skin. Then move through the hottest desert, and not only do your clothes dry, but you must remove some in order to keep cool. The invariance here is you.

LAWS OF MOTION: *See* classical mechanics.

MANY-WORLDS INTERPRETATION: The interpretation taken in this book, also called the parallel universes interpretation. It appears to be consistent with cosmology, relativity theory, quantum mechanics, and even possibly psychology.

MATTER: The stuff of the universe. Matter is said to occupy space, exist in time, and be perceived by the human senses. Modern physics classifies matter according to its atomic and subatomic properties.

MEASUREMENT PROBLEM: The problem unique to quantum mechanics. Whenever a measurement of a physical system takes place, the system "jumps" into one of many possible physical states. So far, there is no way to account for this jump without using some additional concepts. The parallel universes idea arose to solve the measurement problem. Thus, whenever a measurement takes place, the universe splits into as many universes as there are possibilities.

MICROSECOND: A millionth of a second. Light travels about three football stadiums of length in one microsecond. A microsecond compares with one second as one second compares with about eleven and one-half days.

MODULATION: A process where one wave affects another. Usually one of the waves is called the carrier wave and the other is called the information wave. The most common form of modulation is amplitude modulation, used in radio waves. Here the amplitude of the carrier wave changes rather than remaining constant. The changes in the amplitude follow the form contained in the information wave.

MOMENTUM: A measure of matter in motion. A large hunk of matter moving slowly has large momentum because of its mass. A small bit of matter moving quickly has a large momentum because of its speed. In quantum physics, momentum is a primary quality. Thus it is possible for an object to have a well-defined momentum but not have a well-defined mass or speed.

NANOSECOND: A billionth of a second. Light travels about one foot in a nanosecond. One nanosecond compares with a second as one second compares with thirty-two years.

NEWTONIAN PHYSICS: *See* classical mechanics.

NUCLEUS: The core of any atom. The nucleus contains more than 99 percent of the total mass of the whole atom. It consists of subatomic particles grouped in two basic forms: neutrons and protons. The number of protons gives the atom its atomic charge or atomic number. The number of neutrons plus protons gives the atom its atomic mass.

OBSERVER EFFECT: the sudden change in a physical property of matter, particularly at the atomic and subatomic level, when that property is observed. This is measured by the change in the probability of observing that property.

OFFER WAVE: A quantum physical wave of probability that travels forward in time seeking a future event to transact with (*see* transactional interpretation). Until the offer wave is accepted by some future event, neither event can manifest.

PARALLEL UNIVERSES: The idea that instead of just a single universe there exist an infinite number of universes. Matter exists in all of these universes as parallel ghosts.

PHASE: A mathematical function particular to equations representing wave motion. The phase of the wave, like the phase of the moon, is its relationship to some fixed form in space and time. As the phase increases, some physical quantity usually repeats itself periodically. For example, phase can be pictured by looking at a clock face. Every sixty seconds the sweep second hand passes the same point on the dial. When waves are in phase their wave forms match everywhere. When waves are out of phase, their wave forms cancel each other out. Two clocks are in phase when their sweep second hands point to the same respective places on the dials at the same time. Two clocks are out of phase when the hands point in opposite direction (one hand points to twelve while the other points to six, for example). The relative phase between two waves is the same as the angle between two sweep second hands.

PHOTON: The smallest unit of light energy. A photon has measurable properties. It has no electrical charge, no inertial mass—although it is capable of delivering a "punch" of momentum, and a spin twice the magnitude of an electron's.

POWERS OF TEN: A number written as 10^6 simply means the number ten multiplied by itself six times.

PROBABILITY: The mathematical measure of the likelihood of an event occurring. According to the laws of quantum mechanics, probability is a measure of possibilities that must somehow exist simultaneously because these possibilities can effect, or overlap with, each other, thus changing the physical properties of matter.

QUANTUM ELECTRODYNAMICS: The laws of quantum mechanics applied to the study of electrically charged particles. The master equation for this study is Dirac's equation.

QUANTUM MECHANICS: The theory of the behavior of matter and energy, particularly at the level of atoms and subatomic particles. It is nearly impossible to imagine the strangeness of the behavior of matter at this level. An electron in an atom, for example, performs a trick much like the crew aboard the *Enterprise* in the well-known *Star Trek* series, when it "beams" from one energy level to another. It simply jumps from one place to another without going in between.

QUANTUM TUNNELING: The ability of a physical system to tunnel through a physical barrier separating it from the outside world. A classical mechanical system could not do this. This ability arises through the wave properties of the physical system.

QUANTUM WAVE: *See* quantum wave function.

QUANTUM WAVE FUNCTION: A mathematical formula that presents the possibilities of events occurring in the form of a wave pattern distributed through space, much as a wave ripples and flows.

QWIFF: *See* quantum wave function.

RELATIVITIES:

GENERAL RELATIVITY: The theory of the universe that explains the presence of gravity as the distortion of space and time together. If a spacetime distortion is present, there must be matter. To visualize this, imagine a giant rubber sheet that is stretched on a frame. Now imagine placing a lead ball on the sheet so that it distorts the sheet. The sheet is spacetime.

SPECIAL RELATIVITY: A set of rules that enable an observer to calculate what another observer sees when he is moving at a fixed velocity past the first observer. The basis of this theory is the spacetime right triangle, which obeys the formula that states the square of the hypotenuse is the difference of the squares of the time leg and the space leg. Imagine drawing a right triangle on a sheet of paper. The base leg represents movement through space while the altitude leg represents movement through time. If the space leg is longer than the time leg, the hypotenuse represents the actual time observed by a moving observer, whose speed is the ratio of the legs of the triangle. If the space leg is the same length as the time leg, that ratio is unity, corresponding to the velocity of light.

Consequently the hypotenuse represents a length of zero, which means that the mover who moves at light speed experiences no time passing. If the space leg is shorter than the time leg, the mover moves faster than light, and the hypotenuse represents a possible passage backward through time.

SCHRÖDINGER'S CAT: Refers to a poor pussy that has been locked up in a box containing a Rube Goldberg device that will or will not emit cyanide gas depending on the outcome of a single quantum event—the radioactive discharge of an atom. The paradox is: Suppose that the cat is in the box for a period of time wherein the probability is 50 percent that the atom has discharged. If no one looks in the box, is the cat dead or alive?

SELF-CONSISTENCY: The principle that although the universe may be quite bizarre, making no sense when witnessed from a narrow or rigid perspective, it yields predictions that are consistent with each other. In other words, any principle of physics may be strange, but to be consistent it can't make a prediction that would violate its own principle. Thus a chain of logic that leads from an original statement to a sequence of statements that contradict the original statement would not be self-consistent.

SINGULARITY: A point of spacetime where the laws of physics are invalid because all quantities predicted take on infinite values. Singularities are predicted to arise inside of black holes at their very centers.

SPACELIKE: Refers to the interval in space and time between two events. If the distance between the events is longer than the speed of light multiplied by the time interval separating the events, the events are said to be spacelike separated. This means that nothing physical could connect one event with the other because it would need to travel at a speed greater than light speed, which is impossible according to relativity.

SPACETIMES:

SPACETIME ARENA: The vast volume of all the space in the universe and all the time that is. Each point of spacetime is an event marked by its spatial coordinates and its temporal coordinate. Imagine a balloon. Now blow it up and imagine that it keeps inflating ever larger until it occupies all the space there is.

CURVED SPACETIME: The notion that all of space and time together make up a four-dimensional surface that is curved in some way. Imagine a global sphere. Lines of longitude represent time, and lines of latitude represent space. By moving along a line of longitude one passes through the poles. A journey northward after passing through the north pole takes one southward. The north-bound journey is a passage forward in time while the southward journey takes one backward through time.

FLAT SPACETIME: The notion that all of space and time together make up a four-dimensional surface that is as flat as a pancake. Imagine a sheet of paper. Draw both horizontal and vertical lines on it. Horizontal lines represent space while vertical lines represent time.

SUBATOMIC: Smaller than atomic size. A subatomic particle is one that exists or is capable of existing inside of an atom.

SUPERPOSITION: A merging of quantum possibilities that is like the merging of two or more streams running into a river. According to the rules of quantum physics, any physical property is represented by a quantum wave function. This function, because of its wave properties, can be composed of other wave functions, which in turn represent other physical properties. Thus, a quantum wave function that specifies the exact location of an object is composed of quantum wave functions that give all possible momenta to that object. Thus we say that a position wave function is a superposition of momentum wave functions. Similarly a quantum wave function that specifies the momentum of the object is composed of wave functions that give all possible positions of the object.

SUPERSPACE: An imaginary mathematical structure used to envision situations in which there exist more than three dimensions. Physicists who created the concept were attempting to put relativity and quantum physics together in one package. Superspace contains points just as ordinary space does. But each point in superspace marks the location of every object in a whole universe. That is, each point in superspace is a scale model of a whole and distinct universe.

THERMODYNAMICS: The classical laws that govern the behavior of heat in material substances. There are three laws:

1. The conservation of energy—in any physical process the energy can change form but it cannot vanish.
2. Heat cannot spontaneously flow from one body to a warmer body.
3. The temperature absolute zero—where no motion occurs—exists.

TIMELIKE: Refers to the interval in space and time between two events. If the distance between the events is shorter than the speed of light multiplied by the time interval separating the events, the events are said to be timelike separated. This means that anything physical could connect one event with the other, since it could travel from one event to the other at a speed less than light speed, which is possible according to relativity.

TIME-LOOPS: Journeys that loop from present to future and then backward to the present, or journeys that start in the present and go back in time and then return to the present, or any combination thereof. Such loops are not forbidden by the laws of physics, particularly if the starting and ending points are at the same time and space but in separate parallel universes.

Transaction: A connection between an event in the future and an event in the present via quantum echo and offer waves (*see* transactional interpretation). A transaction is complete when an event in the present sends an offer wave to a future event and that future event sends back to the present an echo wave containing a replica of the offer wave. The echo wave's strength depends on the probability that the two events will occur. The stronger the echo, the more likely the occurrence of the sequence of events.

Transactional Interpretation: The interpretation of quantum mechanics, first presented by physicist John G. Cramer, that quantum waves are real and capable of traveling both forward and backward in time. Accordingly, any two physical events, called a *transaction,* require both a forward-in-time traveling wave and a backward-in-time traveling wave to be present in order for the both events to physically manifest. The wave moving forward in time is called an *offer* wave and the event moving backward in time is called an *echo* wave.

Uncertainty Principle: A concept, also called the principle of indeterminism, which reflects the inability to predict the future based on the past or based on the present. It arose from the ideas and thoughts first stated by Werner Heisenberg around 1926 or 1927. It is the cornerstone of quantum physics and provides an understanding of why the world is made of events that cannot be related entirely in terms of cause and effect. It is at the root of all physical matter and may manifest for human beings as doubt and insecurity. If this is so, once it is fully understood, it could create a condition of enlightenment in which the world is seen as an illusion and as a product of mind or consciousness.

Wave-particle Duality: The idea that matter can exist in two guises, wave or particle. As a wave, matter is spread out, distributed throughout space. As a particle, matter is concentrated, occupying only a single point of space at a single time. The duality refers to the impossibility of observing matter in both of its guises simultaneously.

Wigner's Friend: Refers to the Schrödinger cat paradox. Suppose a friend who holds the cage containing the cat decides to look in. He will undoubtedly find a live or a dead cat. But suppose a professor named Wigner has the friend and the caged cat in a closed room. If the professor doesn't look in on the friend, even though the friend has looked at the cat, is the friend in a happy state of mind upon seeing a live cat, or a sad state of mind upon seeing a dead cat? According to quantum rules, until the professor looks, the friend's state cannot be decided.

Wormhole: An opening in space that connects one universe with a parallel universe or else connects two remote regions in a single universe. Wormholes arise inside of black holes, particularly if they are extremely small. An electron could be a wormhole.

BIBLIOGRAPHY

Adams, Douglas. *The Restaurant at the End of the Universe.* New York: Pocket Books, 1980.

Aharonov, Yakir, and Albert, David Z. "Is the Usual Notion of Time Evolution Adequate for Quantum-Mechanical Systems? I." *Physical Review D.* Vol. 29 (1984), p. 223.

_____ "II. Relativistic Considerations." *op. cit.* p. 228.

Aharonov, Yakir, Albert, David Z., and D'Amato, Susan S. "Multiple-Time Properties of Quantum-Mechanical Systems." *Physical Review D.* Vol. 32 (1985), p. 32.

Albert David Z., Aharonov, Yakir, and D'Amato Susan S. "Curious New Statistical Prediction of Quantum Mechanics." *Physical Review Letters.* Vol. 54 (1985), p. 5

Albert, David Z. "How to Take a Photograph of Another Everett World." In *New Techniques and Ideas in Quantum Measurement Theory,* ed. D. M. Greenberger, Vol. 480, Annals of the New York Academy of Sciences, December 30, 1986.

_____. "On Quantum-Mechanical Automata." *Physics Letters.* Vol. 98A (1983), pp. 249–252.

Allman, William F. "Newswatch: The Photon's Split Personality." *Science 86.* June 1986, p. 4.

Asimov, Isaac. *The Gods Themselves.* New York: Doubleday, 1972.

Aspect, Alain, Dalibard, Jean, and Roger, Gerard. "Experimental Test of Bell's Inequalities Using Time-Varying Analyzers." *Physical Review Letters.* Vol. 49, No. 25 (20 December 1982), p. 1804.

Ballentine, L. E., Pearle, P., Walker, E. H., Sachs, M., Koga, T., Gerver, J., and DeWitt, B. "Quantum Mechanics Debate." *Physics Today.* April 1971, p. 36.

Barrow, John D., and Tipler, Frank J. "Closed Universes: Their Future Evolution and Final State." *Monthly Notices Royal Astronomical Society.* Vol. 216 (1985), pp. 395–402.

———. *The Anthropic Cosmological Principle.* New York: Oxford University Press, 1986.

Bass, L. "A Quantum Mechanical Mind-Body Interaction." *Foundations of Physics.* Vol. 5, No. 1 (1975), p. 159.

———. "The Mind of Wigner's Friend." *Hermathena.* No. cxii, 1971.

Bateson, Gregory, *Mind and Nature: A Necessary Unity.* New York: E. P. Dutton, 1979.

Beckenstein, J. "Black Holes and Entropy." *Physical Review.* Vol. D7 (1973), p. 2333.

Bell, J. S. "On the Einstein-Podolsky-Rosen Paradox." *Physics.* Vol. 1 (1964).

Benioff, Paul. "Quantum Mechanical Models of Turing Machines That Dissipate No Energy." *Physical Review Letters.* Vol. 48, No. 23 (7 June 1982), p. 1581.

———. "Quantum Mechanical Hamiltonian Models of Computers." In *New Techniques and Ideas in Quantum Measurement Theory,* ed. D. M. Greenberger, Vol. 480, Annals of the New York Academy of Sciences, December 30, 1986.

Bennett, C. H. "Logical Reversibility of Computation." *IBM Journal of Research and Development.* Vol. 17 (November 1973), p. 525.

Bohm, D. J., Dewdney, C., and Hiley, B. H. "A Quantum Potential Approach to the Wheeler Delayed Choice Experiment." *Nature.* Vol. 315 (1985), p. 294.

Bohr, Niels. "Light and Life." *Atomic Physics and Human Knowledge.* New York: Wiley, 1958.

Borges, Jorges. "The Garden of the Forking Paths." In *Ficciones.* New York: Grove Press, 1962.

Boulle, Pierre. *Time Out of Mind.* New York: Signet Books, 1966.

Bradbury, Ray. *The Silver Locusts: A classic collection of S.F. stories taken from the Martian Chronicles.* London: Corgi Books, 1956.

Brillouin, Leon. *Science and Information Theory.* New York: Academic Press, 1962.

Broyles, A. A. "Derivation of the Probability Rule." *Physical Review D.* Vol. 25, No. 12 (15 June 1982), p. 1.

Bub, Jeffrey. "On the Curious Properties of Quantum Ensembles Which Have Been Both Pre- and Post-selected." *Physical Review Letters* (Preprint).

Carow, Ursula, and Watamura, Satoshi. "Quantum Cosmological Model of the Inflationary Universe." *Physical Review D.* Vol. 32 (1985), p. 1290.

Clark, Ronald W. *Einstein: The Life and Times.* New York: World Publishing Co., 1971.

Cooper, Fred, Haymaker, Richard W., Matsuki, T., and Wang, S. "Quantum Corrections to False Vacuum Decay in the Coleman-Weinberg Potential." *Physical Review D.* Vol. 32 (1985), p. 2049.

Cooper, Leon N., and Van Vechten, Deborah. "On the Interpretation of Measurement within the Quantum Theory." *American Journal of Physics.* Vol. 37, No. 12 (1969), p. 1212.

Costa De Beauregard, O. "S-Matrix, Feynman Zigzag and Einstein Correlation." *Physics Letters.* Vol. 67A, No. 3 (1978), p. 171.

Cramer, John G. "Generalized Absorber Theory and the Einstein-Podolsky-Rosen Paradox." *Physical Review D.* Vol. 22 (1980), p. 362.

______. "The Transactional Interpretation of Quantum Mechanics." *Reviews of Modern Physics.* Vol. 58, No. 3 (July 1986).

______. "An Overview of the Transactional Interpretation of Quantum Mechanics." *International Journal of Theoretical Physics.* Vol. 27, No. 2, pp. 227–236, 1988.

______. "The Quantum Handshake." *Analog.* November 1986.

______. "Alternate Universes II." *Analog.* November 1984.

Cron, O. "Expansion Isotropization During the Inflationary Era." *Physical Review D.* Vol. 32 (1985), p. 2522.

D'Amato, Susan S. "Two-Time States of a Spin-Half Particle in a Uniform Magnetic Field." In *New Techniques and Ideas in Quantum Measurement Theory,* ed D. M Greenberger, Vol. 480, Annals of the New York Academy of Sciences, December 30, 1986.

Davies, Paul. *Superforce.* New York: Simon & Schuster, 1984.

______. *God and the New Physics.* New York: Simon & Schuster, 1983.

______. *Other Worlds.* New York: Simon & Schuster, 1980.

______. *The Edge of Infinity.* New York: Simon & Schuster, 1981.

______. *The Accidental Universe.* Cambridge, England: Cambridge University Press, 1982.

Dennett, Daniel C. *Brainstorms: Philosophical Essays on Mind and Psychology.* Cambridge, Mass.: MIT Press, 1981.

Deutsch, D. "Quantum Theory, the Church-Turing Principle and the Universal Quantum Computer." *Proceedings of the Royal Society of London,* Vol. A 400 (1985), pp. 97–117.

Dewdney, C., Gueret, Ph., Kyprianidis, A., and Vigier, J. P. "Testing Wave-Particle Dualism with Time-Dependent Neutron Interferometry." *Physics Letters.* Vol. 102A, No. 7 (1984), p. 291.

DeWitt, Bryce S. "Quantum Mechanics and Reality." *Physics Today.* Sept. 1970, pp. 30–35.

______. "Quantum Gravity." *Scientific American.* Vol. 249 (1983), p. 112.

DeWitt, Bryce S., and Graham, Neill. *The Many-Worlds Interpretation of Quantum Mechanics.* Princeton, New Jersey: Princeton Univ. Press, 1973.

Dickinson, Terence. "In the Beginning." *Equinox.* Vol. III, No. 15 (May/June 1984, published in Ontario, Canada), p. 56.

Dieks, D. "On the Covariant Description of Wave Function Collapse." *Physics Letters.* Vol. 108A, No. 8 (1985), p. 379.

Dirac, P. A. M. *The Principles of Quantum Mechanics.* London: Oxford University Press, 1958.

Duncan, Ronald, and Weston-Smith, Miranda (eds.). *The Encylopaedia of Ignorance: Everything you ever wanted to know about the unknown.* Elmsford, New York: Pergamon Press, Ltd., 1977.

Einstein, A., Lorentz, H. A., Weyl, H., and Minkowski, H. *The Principle of Relativity.* New York: Dover Publications, 1923.

Einstein, Albert, Podolsky, Boris, and Rosen, Nathen. "Can the Quantum-Mechanical Description of Physical Reality Be Considered Complete?" *Physical Review.* Vol. 47 (1935), p. 777

Einstein, A. and Rosen, N. "Particle Problem in the General Theory of Relativity." *Physical Review.* Vol. 48 (1935), pp. 73—77.

Elvee, Richard Q. (ed.). *Mind in Nature, with contributions by Granit, Pannenberg, Popper, Rorty, Wheeler, and Wigner. Nobel Conference XVII.* San Francisco: Harper & Row, 1982.

Epstein, Lewis C. *Relativity Visualized.* San Francisco: Insight Press 1983.

Evans, Mark, and McCarthy, James G. "Quantum Mechanics of Inflation." *Physical Review D.* Vol. 31 (1985), p. 1799

Franson, J. D. "Extension of the Einstein-Podolsky-Rosen Paradox and Bell's Theorem," *Physical Review D.* Vol. 26, No. 4 (1982).

———. "An Experimental Test of Locality Using a Single Photon Interferometer." In *New Techniques and Ideas in Quantum Measurement Theory,* ed. D. M. Greenberger, Vol. 480, Annals of the New York Academy of Sciences, December 30, 1986.

Freedman, Daniel Z., and Nieuwenhuizen, Peter van. "The Hidden Dimensions of Spacetime." *Scientific American.* Vol. 252 (1985), p. 74.

Feynman, Richard P. "Simulating Physics with Computers." *International Journal of Theoretical Physics.* Vol. 21, Nos. 6/7 (1982), pp. 467–488.

Gautreau, Ronald. "Physically Incomplete Extensions of the Schwarzschild Metric." *New Jersey Institute of Technology Report* (1978).

———. "Alternate Views to Black Holes." *op. cit.* (Sept. 1982).

———. "Imbedding a Schwartzschild Mass into Cosmology." *op. cit.* (1982).

———. "A New View of the Big Bang." *op. cit.* (1982).

———. "Curvature Coordinates in Cosmology." *op. cit.* (no date given).

———. "On the Light Cone Inside the Schwartzschild Radius." *op. cit.* (no date given).

———. "On Kruskal-Novikov Co-ordinate Systems." *Il Nuovo Cimento.* Vol. 56 (1980), p. 49.

———. "Geodesic Coordinates in the de Sitter Universe." *Physical Review D.* Vol. 27 (1983), p. 764.

Gautreau, Ronald, and Hoffman, Banesh. "The Schwarzschild Radial Coordinate as a Measure of Proper Distance." *Physical Review D.* Vol. 17 (1978), p. 2552.

Gott III, J. Richard. "A Time-Symmetric, Matter, Antimatter, Tachyon Cosmology." *The Astrophysical Journal.* Vol. 187 (1974), p. 1.

Grangier, P., Roger, G., and Aspect, A. "A New Light on Single Photon Interferences." In *New Techniques and Ideas in Quantum Measurement Theory,* ed. D. M. Greenberger, Vol. 480, Annals of the New York Academy of Sciences, December 30, 1986.

Grangier, Phillipe, Aspect, Alain, and Vigue, Jacques. "Quantum Interference Effect for Two Atoms Radiating a Single Photon." *Physical Review Letters.* Vol. 54 (1985), p. 418.

Greenberger, Daniel M., and Yasin, Allaine. "The Haunted Measurement in Quantum Theory." In *New Techniques and Ideas in Quantum Measurement Theory,* ed. D. M. Greenberger, Vol. 480, Annals of the New York Academy of Sciences, December 30, 1986.

Griffiths, Robert B. "Making Consistent Inferences from Quantum Measurements." In *New Techniques and Ideas in Quantum Measurement Theory,* ed. D. M. Greenberger, Vol. 480, Annals of the New York Academy of Sciences, December 30, 1986.

Guth, Alan H., and Pi, So-Young. "Quantum Mechanics of the Scaler Field in the New Inflationary Universe." *Physical Review D.* Vol. 32 (1985), p. 1899.

Guth, Alan H. "Inflation Universe: A Possible Solution to the Horizon and Flatness Problem." *Physical Review D.* Vol. 23 (1981), p. 347.

Guth, Alan H., and Steinhardt, Paul J. "The Inflationary Universe." *Scientific American.* Vol. 250 (1984), p. 116.

Hanggi, Peter. "Macroscopic Quantum Tunneling at Finite Temperatures." In *New Techniques and Ideas in Quantum Measurement Theory,* ed. D. M. Greenberger, Vol. 480, Annals of the New York Academy of Sciences, December 30, 1986.

Hartle, J. B., and Hawking, S. W. "Wave Function of the Universe." *Physical Review D.* Vol. 28 (1983), p. 2960.

Hawking, Stephen. *A Brief History of Time.* New York: Bantam Publishing Company, 1988.

Hawking, S. W. "Arrow of Time in Cosmology." *Physical Review D.* Vol. 32 (1985), p. 2489.

Hellmuth, T., Zajonc, Arthur C., and Walther, H. "Realizations of Delayed Choice Experiments." In *New Techniques and Ideas in Quantum Measurement Theory,* ed. D. M. Greenberger, Vol. 480, Annals of the New York Academy of Sciences, December 30, 1986.

Hoyle, Fred. *The Intelligent Universe.* New York: Holt, Rinehart & Winston, 1983.

______. "The Universe: Past and Present Reflections," *Preprint Series No. 70, Astrophysics and Relativity.* Cardiff, United Kingdom: Department of Applied Mathematics and Astronomy, University College, May 1981.

Jahn, Robert G. "The Persistent Paradox of Psychic Phenomena: An Engineering Perspective." *Proceedings of the Institute of Electrical and Electronics Engineers (IEEE).* Vol. 70, No. 2 (Feb. 1982), p. 136.

Jammer, Max. *The Philosophy of Quantum Mechanics.* New York: John

Wiley & Sons, 1974.

———. *The Conceptual Development of Quantum Mechanics.* New York: McGraw-Hill Book Co., 1966.

Joos, E. "Quantum Theory and the Appearance of a Classical World." In *New Techniques and Ideas in Quantum Measurement Theory,* ed. D. M. Greenberger, Vol. 480, Annals of the New York Academy of Sciences, December 30, 1986.

Kazama, Yoichi, and Nakayama, Ryuichi. "Wave Packet in Quantum Cosmology." *Physical Review D.* Vol. 32 (1985), p. 2500.

Kaufmann, W. III. *The Cosmic Frontiers of General Relativity.* Boston: Little, Brown, 1979.

Kerr, Roy P. "Gravitational Field of a Spinning Mass as an Example of Algebraically Special Metrics." *Physical Review Letters.* Vol. 11 (1963), pp. 237–38.

Leggett, A. J. "Quantum Mechanics and Realism at the Macroscopic Level: Is an Experimental Discrimination Feasible?" In *New Techniques and Ideas in Quantum Measurement Theory,* ed. D. M. Greenberger, Vol. 480, Annals of the New York Academy of Sciences, December 30, 1986.

Le Guin, Ursula. *The Lathe of Heaven.* New York: Avon, 1973.

Malin, S. "The Meaning and Significance of Quantum States." An invited paper for the *Nathan Rosen Issue of Foundations of Physics* on the occasion of Prof. Rosen's 75th birthday. Preprint from Dept. of Physics & Astronomy, Colgate University, New York.

Mazenko, Gene F., Unruh, William G., and Wald, R. M. "Does a Phase Transition in the Early Universe Produce the Conditions Needed for Inflation?" *Physical Review D.* Vol. 31 (1985), p. 273.

McCorduck, Pamela. *Machines Who Think.* San Francisco: W. H. Freeman & Co., 1979.

Mermin, N. D. "Bringing Home the Atomic World: Quantum Mysteries for Anybody." *American Journal of Physics.* Vol. 49 (1981), p. 940.

Misner, C., Thorne, K., and Wheeler, J. A. *Gravitation.* San Francisco: Freeman, 1973.

Morris. W. (ed). *The American Heritage Dictionary of the English Language.* Boston: American Heritage Publishing Co. and Houghton Mifflin Co., 1969.

Moss, I. G. "Black-Hole Bubbles." *Physical Review D.* Vol. 32 (1985), p. 1333.

Overbye, Dennis. "The Shadow Universe." *Discover Magazine.* May 1985.

Page, Don N. "Will Entropy Decrease If the Universe Recollapses?" *Physical Review D.* Vol. 32 (1985), p. 2496.

Rohrlich, Fritz. "Reality and Quantum Mechanics." In *New Techniques and Ideas in Quantum Measurement Theory,* ed. D. M. Greenberger, Vol. 480, Annals of the New York Academy of Sciences, December 30, 1986.

Stapp, Henry P. "Einstein Locality, EPR Locality, and the Significance for Science of the Nonlocal Character of Quantum Theory." *Lawrence*

Berkeley Laboratory Report–20094, Oct. 1985.
Sutherland, R. I. "A Corollary to Bell's Theorem." *Il Nuovo Cimento*. Vol. 88B, No. 2 (1985), p. 114.
Thomsen, Dietrick E. "A Knowing Universe Seeking to be Known." *Science News*. Vol. 123 (Feb. 19, 1983), p. 124.
______. "The Quantum Universe: A Zero-Point Fluctuation?" *op. cit.* Vol. 128 (1985), p. 72.
______. "Going Bohr's Way in Physics." *op. cit.* Vol. 129 (1986), p. 26.
______. "Holism and Particlism in Physics." *op. cit.* Vol. 129 (1986), p. 70.
______. "Quanta at Large: 101 Things to Do with Schrödinger's Cat." *op. cit.* Vol. 129 (1986), p. 87.
______ "Notes of an Ex-Physics Student." *op. cit.* Vol. 129 (1986), p. 141.
Tipler, Frank J. "Rotating Cylinders and the Possibility of Global Causality Violation." *Physical Review D*. Vol. 9 (1974), p. 2203.
______. "Penrose Diagrams for the Einstein, Eddington-Lemaitre-Bondi, and anti-de Sitter Universes." *Journal of Mathematical Physics*. Vol. 27, No. 2 (February 1986), p. 559.
______. "Cosmological Limits on Computation." *International Journal of Theoretical Physics*. Vol. 25, No. 6, 1986.
______. "Interpreting the Wave Function of the Universe." *Physics Reports*. Vol. 137 (May 1986), p. 4.
Toben, Bob, and Wolf, Fred Alan. *Space-Time and Beyond*. The New Edition. New York: E. P. Dutton, 1982. Also New York: Bantam Edition, 1983.
Unruh, W. G. "Notes on Black Hole Evaporation." *Physical Review D*. Vol. 14 (1976), p. 870.
Vilenkin, Alexander. "Classical and Quantum Cosmology of the Starobinsky Inflationary Model." *Physical Review D*. Vol. 32 (1985), p. 2511.
Vonnegut, Kurt. *Slaughterhouse Five*. New York: Dell Publishing Co., 1969.
Weinberg, Stephen. *The First Three Minutes*. New York: Basic Books, 1977.
Wheeler, John A. "The Mystery and the Message of the Quantum." Presentation at the *Joint Annual Meeting of the American Physical Society and the American Association of Physics Teachers*. Jan. 1984.
______. "Delayed-Choice Experiments and the Bohr-Einstein Dialogue." Paper read at a joint Meeting of The American Philosophical Society and the Royal Society, June 5, 1980, London. Library of Congress catalog card number 80-70995, 1980.
______. "Beyond the Black Hole." In *Some Strangeness in the Proportion: A Centennial Symposium to Celebrate the Achievements of Albert Einstein*. Edited by Harry Woolf. Reading, Mass.: Addison-Wesley, 1980, p. 341.
Wolf, Fred Alan. *Taking the Quantum Leap: The New Physics for Nonscientists*. San Francisco: Harper & Row, 1981.
______. *Star Wave: Mind, Consciousness, and Quantum Physics*. New York: Macmillan, 1984.
______. *The Body Quantum: The New Physics of Body, Mind, and Health*. New York: Macmillan, 1986.
Woo, C. H. "Consciousness and Quantum Interference—An Experimental

Approach." *Foundations of Physics.* Vol. 11, Nos. 11/12 (1981).
Wooters, W. K. "Statistical Distances and Hilbert Space." *Physical Review D.* Vol. 23 (1981), p. 357.
Zurek, W. H., and Thorne, Kip S. "Statistical Mechanical Origin of the Entropy of a Rotating Charged Black Hole." *Physical Review Letters.* Vol. 54 (1985), p. 2171.

INDEX

ABOUT THE AUTHOR

Fred Alan Wolf is a physicist and the author of five books, including *Taking the Quantum Leap,* which won an American Book Award. He has taught at San Diego State University, the University of Paris, the Hebrew University of Jerusalem, and other colleges and universities in the United States and abroad. He also lectures and consults worldwide.